AF368958

Assessing the Effects of Multiple Stressors on Aquatic Systems across Temporal and Spatial Scales: From Measurement to Management

Assessing the Effects of Multiple Stressors on Aquatic Systems across Temporal and Spatial Scales: From Measurement to Management

Editors

Pedro Segurado
Paulo Branco
Maria Teresa Ferreira

MDPI • Basel • Beijing • Wuhan • Barcelona • Belgrade • Manchester • Tokyo • Cluj • Tianjin

Editors
Pedro Segurado
Forest Research Centre,
School of Agriculture,
University of Lisbon
Portugal

Paulo Branco
Forest Research Centre,
School of Agriculture,
University of Lisbon
Portugal

Maria Teresa Ferreira
Forest Research Centre,
School of Agriculture,
University of Lisbon
Portugal

Editorial Office
MDPI
St. Alban-Anlage 66
4052 Basel, Switzerland

This is a reprint of articles from the Special Issue published online in the open access journal *Water* (ISSN 2073-4441) (available at: https://www.mdpi.com/journal/water/special_issues/Stressors_Aquatic_Systems).

For citation purposes, cite each article independently as indicated on the article page online and as indicated below:

LastName, A.A.; LastName, B.B.; LastName, C.C. Article Title. *Journal Name* **Year**, *Volume Number*, Page Range.

ISBN 978-3-0365-4199-0 (Hbk)
ISBN 978-3-0365-4200-3 (PDF)

Contents

Editorial

Assessing the Effects of Multiple Stressors on Aquatic Systems across Temporal and Spatial Scales: From Measurement to Management

Pedro Segurado *, Teresa Ferreira and Paulo Branco

Forest Research Centre, School of Agronomy, University of Lisbon, Tapada da Ajuda, 1349-017 Lisboa, Portugal; terferreira@isa.ulisboa.pt (T.F.); pjbranco@isa.ulisboa.pt (P.B.)
* Correspondence: psegurado@isa.ulisboa.pt

Citation: Segurado, P.; Ferreira, T.; Branco, P. Assessing the Effects of Multiple Stressors on Aquatic Systems across Temporal and Spatial Scales: From Measurement to Management. *Water* **2021**, *13*, 3549. https://doi.org/10.3390/w13243549

Received: 22 November 2021
Accepted: 8 December 2021
Published: 11 December 2021

Publisher's Note: MDPI stays neutral with regard to jurisdictional claims in published maps and institutional affiliations.

Freshwater habitats are home to a disproportionately high biodiversity, given the total area they cover worldwide, hosting 10% of all species while occupying less than 1% of the Earth's surface [1]. These habitats have long been affected by a wide range of co-occurring environmental stressors that disrupt freshwater biodiversity and ecosystem functioning, hence menacing the provision of ecosystem services that are vital to human well-being, including water supply and food security [2–4]. Despite the increasing governance awareness in many parts of the world, as evidenced by the implementation of legislation, policies and regulations such as the Clean Water Act in USA, the Water Framework Directive in the EU and the Water Act in Australia, freshwater ecosystems are far from recovery and most likely to be exposed to new stressors, given the escalade of emerging threats [4–7]. This is evidenced by the also disproportionate percentage of endangered fauna and flora: of the freshwater dependent species so far assessed in the IUCN Red List, 27% are classified as threatened with extinction [8].

Despite the ever-increasing body of research on multiple stressor effects, the efforts made so far to apply the acquired knowledge on concrete management actions on freshwater habitats, including environmental restoration and protection programmes, have resulted in improvements that fall well below the expectations [9,10]. This overall inefficiency is often claimed to be the consequence of knowledge gaps on how individual stressors act in concert [10,11], especially by limiting our capacity to generalise, and therefore predict, ecological responses under strategies involving single stressor reduction [12]. At the same time, these multiple stressors act simultaneously at different spatial and time scales, with their effects being susceptible to vary with climate changes [13,14], local natural conditions [15] and spatial scale [16]. There are still many challenges to implement efficient management practices, such as by improving the understanding of the mechanisms underlying stressor interactions, adapting monitoring programmes to new evidence on the relationships between multiple-stressor interactions and ecological responses, and shifting the focus from ecosystem degradation pathways—which has been so far the main focus of multiple stressor research—to the processes that govern recovery [5,12].

By acknowledging these important research challenges, in this Special Issue we proposed to bring together research advances on the topic of stressors interplay across spatial and temporal scales and its consequences for management of aquatic systems. This Special Issue gathers six very diverse publications, including one review and five research articles, from four continents: Europe [17–19], Oceania [20], Africa [21] and North America [22]. The research articles include studies focused on fish [17,18], macroinvertebrates [19,20], and phytoplankton [17], and are based either on field data [17,19,20], historical data [18] and interviews/expert knowledge [21]. The analysed stressors include physical and chemical barriers [18], hydromorphology, land use and water quality [19], nutrient enrichment and temperature [17], livestock farming and flow reduction [20], water level and temperature [21], and diffuse pollution from croplands and rangelands [22].

In Europe, Le Pichon and her colleagues [18] took advantage of a multitude of historical sources to evaluate the historical evolution of the potential cumulative impacts of physical (weirs, locks, and hydropower plants) and chemical (dissolved oxygen) barriers on the habitat accessibility of diadromous fish species in the Lower Seine River, France. They thoroughly combined historical data sources, such as engineering projects, navigation maps, records of monthly average dissolved oxygen, with knowledge on engineering sciences and fish ecology to develop a least-cost-based connectivity model for three migratory species with distinct migratory behaviours (Atlantic salmon, allis shad, and sea lamprey) at three time periods (1900s, 1970s and 2000s). They found that accessibility, as measured by effective functional distances, varied with fish migration behaviour, time period, and the level of tolerance to low dissolved oxygen. The highest disruptions of ecological connectivity were estimated for the 1970s, corresponding to the post-war industrialization period, due to the joint effect of wide hypoxic river segments together with the installation of impassable navigation weirs (in which many fish passes were only later installed). Several management recommendations are discussed in light of the main findings, namely the importance of controlling chemical water quality while maintaining or increasing the effectiveness of fish passages.

In a study also conducted in France, Bouraï et al. [17], based on a biomonitoring dataset comprising 204 lakes, investigated how two major stressors in lakes related to climate changes—nutrient enrichment and temperature increase—interact in their impacts on the community structure of two biological groups occupying extreme positions of lake food webs (phytoplankton and fish). They modelled the effects of these two stressors on different community metrics related to abundance, composition, size structure, and size spectra, taking also into account the natural environmental variability. Among the significantly responsive metrics, the majority (four metrics) were affected by a single stressor and only fish-based metrics were affected by more than one stressor: one—the number of individuals caught per sampling unit—responded additively to temperature and eutrophication, and two—the perch/roach biomass ratio and the average fish size—were impacted by antagonistic interactions, in which one stressor was found to attenuate the effect of the other. They also stress that modelled patterns for stressor combinations outside the range of existing conditions in the dataset (for example lakes that are simultaneously cold and eutrophic) are due to statistical artifacts.

Heading slightly eastwards, Urbanic and his colleagues [19] examined the single and joint effects of natural factors and three major stressor groups—hydromorphology, land use, and water quality—on the benthic macroinvertebrate community structure in five large rivers of Slovenia and Croatia, based on field data collected over a wide range of environmental conditions, from near-natural sites to heavily altered rivers. This study represents a huge challenge, since biological sampling is very demanding in large rivers, as they typically are affected by a complex combination of stressors that in great measure result from the cumulative conditions that converge from the upstream tributaries. Their analyses were based on multivariate constrained ordination techniques to extract the major community gradients as the response to stressors combinations and river typology, followed by a variation partitioning approach. They found that the pure contribution of hydromorphological, land use, and water quality gradients dominated over both river typology and shared effects in structuring large river macroinvertebrate assemblages. They claim that the dominance of pure stressor contributions found in this study will help managers to better understand the ecological changes that large rivers have experienced in the past and to predict how ecological status and ecosystem services will evolve under future environmental changes.

Moving to the extreme Southeast of the globe, Lange and her colleagues [20], conducted an innovative study that use nitrogen stable isotope values ($\delta_{15}N$) of three invertebrate grazers as potential indicators of land-use intensification to investigate the combined effects of farming intensity and flow reduction in the Manuherikia River catchment in South Island, New Zealand. They found that variations of $\delta_{15}N$ values along stressor gradients were not consistent among the targeted primary consumers. The larvae of mayfly *Deleatidium* spp. belonged to the only species for which the δ_{15}N values showed the expected positive relationship with sheep/beef farming intensity, which was found to interact antagonistically with flow reduction, i.e., the latter attenuated the effect of former stressor. The positive response of δ_{15}N values to farming intensity was attributed to processes such as inputs of industrial fertilizers, animal waste products and nitrogen transformation processes (e.g., denitrification and ammonia volatilization in agricultural soils and streams). The antagonistic effect may arise when nitrogen input under flow reduction decreases to such an extent that weakens the positive effect of increased farming intensity. In contrast, the δ_{15}N of the two analyzed snail species either showed a positive response to farming intensity (*Physella* spp.) or no significant response (*Potamopyrgus* spp.). The differences found in consumer δ_{15}N values were attributed to the likely ingestion of different components of the periphytic community, probably driven by differences in microhabitat use, something the authors recommend to be investigated in future studies. The authors also conclude that the mayfly *Deleatidium* spp. is likely well-suited as a bioindicator in stable isotope studies on agricultural impacts in New Zealand, given its high density, widespread distribution, strict dietary preference, and the clear response of δ_{15}N values to farming intensity.

In the Sahel region of Africa, Sanon and colleagues [21] conducted an ambitious study that fills important knowledge gaps in freshwater ecosystems of semi-arid and resource-poor countries. Their study aimed at understanding the joint effects of multiple socio-ecological stressors on the ecological integrity of aquatic ecosystems in the Nakambe River (or White Volta), in Burkina Faso, to support and improve fishery management efforts under ongoing climate changes. For that purpose, they gathered a wide range of qualitative data from literature reviews, interviews and strategic simulations (i.e., interactive participatory methods involving experts and stakeholders) as multiple lines of evidence across a Drivers–Pressure–State–Impact–Response (DPSIR) framework. They show how fish productivity, abundance, and average body size, and consequently social well-being indicators such as food and nutrition security, are affected by human impacts as well as climate change effects, namely on water level and surface water temperature. These impacts are further exacerbated by the ongoing nutrition transition towards a greater demand on proteins. They recommend a series of policy responses such as increasing measures for family planning, encouraging and empowering the participation of the different actors to reinforce fisheries regulation and develop the provision of alternative livelihood, such as aquaculture. These measures would help achieving the sustainable management of aquatic ecosystems, promoting the recovery of fish stocks in natural ecosystems, reducing fishermen's vulnerability and preventing further poverty and food insecurity.

Finally, in the USA, a country that, despite having pioneered environmental legislation on freshwaters with one of the most worldwide influential environmental laws, the Clean Water Act from the 1970s, still has a long way to go on the environmental protection of freshwaters according to the literature review conducted by Hughes and Vadas [22]. They focus their review on the effects of croplands and rangelands on freshwaters, by posing a series of questions and presenting a list of case studies. Only 26–30% of the entire stream/river length of conterminous USA streams and rivers were estimated to be in good conditions. Agriculture has been pointed out as a main driver of water quality impairment in USA surface waters and in their review, Hughes and Vadas give some examples where the prevalence of multiple stressors contexts related with a range of cropland and rangeland activities support this view. They summarize the main outputs of research case studies on best management practices and livestock exclosures to provide a general picture of how multiple stressors are affecting biotic indicators and list a series of management challenges for improving the biotic condition of streams draining croplands and rangelands. They end their review by discussing management and governance recommendations to mitigate the problems of diffuse pollution from croplands and rangelands, such as the need to reinforce the focus on biotic and groundwater variables.

Overall, the articles included in this Special Issue provide a representative view of how multiple stressors in freshwaters, notably river fragmentation, nutrient enrichment, flow reduction and surface temperature, are currently being addressed by researchers, managers and decision-makers. Despite pointing out important limitations and challenges that need to be faced to tackle multiple stressor effects on freshwaters, they all end up showing some optimistic perspectives for the future of freshwater ecosystems, either by referring to promising outcomes of previous and ongoing management and protection measures [18,22], demonstrating some benefits from technical advances [20], disentangling multiple stressor effects that will ease management planning [17,19] and, last but not least, indicating how international cooperation between researchers and local stakeholders of undeveloped countries with serious natural resource limitations might contribute to the environmental sustainability of their freshwater ecosystems, as well as the services they provide [21].

Funding: Forest Research Centre (CEF) is a research unit funded by Fundação para a Ciência e a Tecnologia I.P. (FCT), Portugal (UID/AGR/00239/2019), and is part of the cluster Associated Laboratory TERRA, LA/P/0092/2020. This article was partially funded by a project granted to P Segurado by FCT under the IF Researcher Programme (IF/01304/2015). P Branco has been financed by national funds via FCT—Fundação para a Ciência e a Tecnologia, I.P., under "Norma Transitória–DL 57/2016/CP1382/CT0020".

Acknowledgments: The guest editors of this Special Issue express their thanks to the authors that joined the call, as well as to the reviewers for their valuable contribution to improve the original manuscripts.

Conflicts of Interest: The authors have no conflict of interest to declare.

References

1. Strayer, D.L.; Dudgeon, D. Freshwater biodiversity conservation: Recent progress and future challenges. *J. N. Am. Benthol. Soc.* **2010**, *29*, 344–358. [CrossRef]
2. Vörösmarty, C.J.; McIntyre, P.B.; Gessner, M.O.; Dudgeon, D.; Prusevich, A.; Green, P.; Glidden, S.; Bunn, S.E.; Sullivan, C.A.; Liermann, C.R.; et al. Global threats to human water security and river biodiversity. *Nature* **2010**, *467*, 555–561. [CrossRef] [PubMed]
3. Carpenter, S.R.; Stanley, E.H.; Vander Zanden, M.J. State of the world's freshwater ecosystems: Physical, chemical, and biological changes. *Annu. Rev. Environ. Resour.* **2011**, *36*, 75–99. [CrossRef]
4. Reid, A.J.; Carlson, A.K.; Creed, I.F.; Eliason, E.J.; Gell, P.A.; Johnson, P.T.J.; Kidd, K.A.; MacCormack, T.J.; Olden, J.D.; Ormerod, S.J.; et al. Emerging threats and persistent conservation challenges for freshwater biodiversity. *Biol. Rev.* **2019**, *94*, 849–873. [CrossRef]
5. Craig, L.S.; Olden, J.D.; Arthington, A.H.; Entrekin, S.; Hawkins, C.P.; Kelly, J.J.; Kennedy, T.A.; Maitland, B.M.; Rosi, E.J.; Roy, A.H. Meeting the challenge of interacting threats in freshwater ecosystems: A call to scientists and managers. *Elem. Sci. Anthr.* **2017**, *5*, 72. [CrossRef]
6. Dudgeon, D. Multiple threats imperil freshwater biodiversity in the Anthropocene. *Curr. Biol.* **2019**, *29*, 960–967. [CrossRef]
7. He, F.; Zarfl, C.; Bremerich, V.; David, J.N.; Hogan, Z.; Kalinkat, G.; Tockner, K.; Jähnig, S.C. The global decline of freshwater megafauna. *Glob. Chang. Biol.* **2019**, *25*, 3883–3892. [CrossRef] [PubMed]
8. Tickner, D.; Opperman, J.; Abell, R.; Acreman, M.; Arthington, A.H.; Bunn, S.E.; Cooke, S.J.; Dalton, J.; Darwall, W.; Edwards, G.; et al. Bending the Curve of Global Freshwater Biodiversity Loss: An Emergency Recovery Plan. *BioScience* **2020**, *70*, 330–342. [CrossRef] [PubMed]
9. Lemm, J.U.; Venohr, M.; Globevnik, L.; Stefanidis, K.; Panagopoulos, Y.; van Gils, J.; Posthuma, L.; Kristensen, P.; Feld, C.K.; Mahnkopf, J.; et al. Multiple stressors determine river ecological status at the European scale: Towards an integrated understanding of river status deterioration. *Glob. Chang. Biol.* **2021**, *27*, 1962–1975. [CrossRef]
10. Vigiak, O.; Udias, A.; Pistocchi, A.; Zanni, M.; Aloe, A.; Grizzetti, B. Probability maps of anthropogenic impacts affecting ecological status in European rivers. *Ecol. Indic.* **2021**, *126*, 107684. [CrossRef] [PubMed]
11. Carvalho, L.; Mackay, E.B.; Cardoso, A.C.; Baattrup-Pedersen, A.; Birk, S.; Blackstock, K.L.; Borics, G.; Borja, A.; Feld, C.K.; Ferreira, M.T.; et al. Protecting and restoring Europe's waters: An analysis of the future development needs of the Water Framework Directive. *Sci. Total Environ.* **2019**, *658*, 1228–1238. [CrossRef]
12. Spears, B.; Chapman, D.; Carvalho, L.; Feld, C.; Gessner, M.; Piggott, J.; Banin, L.; Gutiérrez-Cánovas, C.; Solheim, A.L.; Richardson, J.; et al. Making Waves. Bridging theory and practice towards multiple stressor management in freshwater ecosystems. *Water Res.* **2021**, *196*, 116981. [CrossRef]

13. Nõges, P.; Van De Bund, W.; Cardoso, A.C.; Heiskanen, A.S. Impact of climatic variability on parameters used in typology and ecological quality assessment of surface waters-Implications on the Water Framework Directive. *Hydrobiologia* **2007**, *584*, 373–379. [CrossRef]

14. Bruno, D.; Belmar, O.; Maire, A.; Morel, A.; Dumont, B.; Datry, T. Structural and functional responses of invertebrate communities to climate change and flow regulation in alpine catchments. *Glob. Chang. Biol.* **2019**, *25*, 1612–1628. [CrossRef]

15. Brucet, S.; Pédron, S.; Mehner, T.; Lauridsen, T.L.; Argillier, C.; Winfield, I.J.; Volta, P.; Emmrich, M.; Hesthagen, T.; Holmgren, K.; et al. Fish diversity in European lakes: Geographical factors dominate over anthropogenic pressures. *Freshwater Biol.* **2013**, *58*, 1779–1793. [CrossRef]

16. Birk, S.; Chapman, D.; Carvalho, L.; Spears, B.M.; Andersen, H.E.; Argillier, C.; Auer, S.; Baattrup-Pedersen, A.; Banin, L.; Beklioğlu, M.; et al. Impacts of multiple stressors on freshwater biota across spatial scales and ecosystems. *Nat. Ecol. Evol.* **2020**, *4*, 1060–1068. [CrossRef]

17. Bouraï, L.; Logez, M.; Laplace-Treyture, C.; Argillier, C. How Do Eutrophication and Temperature Interact to Shape the Community Structures of Phytoplankton and Fish in Lakes? *Water* **2020**, *12*, 779. [CrossRef]

18. Le Pichon, C.; Lestel, L.; Courson, E.; Merg, M.; Tales, E.; Belliard, J. Historical Changes in the Ecological Connectivity of the Seine River for Fish: A Focus on Physical and Chemical Barriers Since the Mid-19th Century. *Water* **2020**, *12*, 1352. [CrossRef]

19. Urbanič, G.; Mihaljević, Z.; Petkovska, V.; Pavlin Urbanič, M. Disentangling the Effects of Multiple Stressors on Large Rivers Using Benthic Invertebrates—A Study of Southeastern European Large Rivers with Implications for Management. *Water* **2020**, *12*, 621. [CrossRef]

20. Lange, K.; Townsend, C.; Matthaei, C. Inconsistent Relationships of Primary Consumer N Stable Isotope Values to Gradients of Sheep/Beef Farming Intensity and Flow Reduction in Streams. *Water* **2019**, *11*, 2239. [CrossRef]

21. Sanon, V.; Toé, P.; Caballer Revenga, J.; El Bilali, H.; Hundscheid, L.; Kulakowska, M.; Magnuszewski, P.; Meulenbroek, P.; Paillaugue, J.; Sendzimir, J.; et al. Multiple-Line Identification of Socio-Ecological Stressors Affecting Aquatic Ecosystems in Semi-Arid Countries: Implications for Sustainable Management of Fisheries in Sub-Saharan Africa. *Water* **2020**, *12*, 1518. [CrossRef]

22. Hughes, R.; Vadas, R. Agricultural Effects on Streams and Rivers: A Western USA Focus. *Water* **2021**, *13*, 1901. [CrossRef]

Article

Historical Changes in the Ecological Connectivity of the Seine River for Fish: A Focus on Physical and Chemical Barriers Since the Mid-19th Century

Céline Le Pichon [1,*], Laurence Lestel [2], Emeric Courson [1], Marie-Line Merg [1], Evelyne Tales [1] and Jérôme Belliard [1]

[1]   INRAE, Université Paris-Saclay, HYCAR, 1 rue Pierre-Gilles de Gennes, CS 10030, 92761 Antony, France; emeric.courson@inrae.fr (E.C.); marie-line.merg@inrae.fr (M.-L.M.); evelyne.tales@inrae.fr (E.T.); jerome.belliard@inrae.fr (J.B.)

[2]   Faculté des sciences, Sorbonne Université, UMR METIS, 4, place Jussieu, 75252 Paris CEDEX 05, France; laurence.lestel@sorbonne-universite.fr

*   Correspondence: celine.lepichon@inrae.fr

Received: 12 February 2020; Accepted: 29 April 2020; Published: 10 May 2020

Abstract: To understand the long-term fate of fish assemblages in the context of global change and to design efficient restoration measures in river management, it is essential to consider the historical component of these ecosystems. The human-impacted Seine River Basin is a relevant case that has experienced the extinction of diadromous fishes over the last two centuries and has recently witnessed the recolonization of some species. One key issue is to understand the historical evolution of habitat accessibility for these migratory species. Thanks to the unique availability of historical, mainly hand-written sources of multiple types (river engineering projects, navigation maps, paper-based databases on oxygen, etc.), we documented and integrated, in a geographic information system-based database, the changes to physical and chemical barriers in the Seine River from the sea to Paris for three time periods (1900s, 1970s, and 2010s). The potential impact of these changes on the runs of three migratory species that have different migratory behaviors—Atlantic salmon, allis shad, and sea lamprey—was evaluated by ecological connectivity modeling, using a least-cost approach that integrates distance, costs, and risks related to barriers. We found that accessibility was contrasted between species, emphasizing the crucial role of the migration type, period, and level of tolerance to low dissolved oxygen values. The highest disruption of ecological connectivity was visible in the 1970s, when the effects of large hypoxic areas were compounded by those of impassable navigation weirs (i.e., without fish passes). As the approach was able to reveal the relative contribution of physical and chemical barriers on overall functional connectivity, it may constitute a model work in assessing the functioning of large river ecosystems.

Keywords: least-cost modeling; longitudinal connectivity; dissolved oxygen; historical data; functional distance; migratory fish; fish passes; navigation weir

---

1. Introduction

Riverine hydrosystems are highly complex socio-ecological systems, reflecting a long interwoven history between rivers and societies. They are structurally complex, biodiverse, and productive due to their dendritic structure, connectivity with adjacent water bodies, and multiple relationships with terrestrial and marine ecosystems [1]. The varied and increasing use of streams and rivers by human societies through time has drastically modified riverscapes and, consequently, ecosystem functions and biodiversity. Today, large riverine hydrosystems are among the most threatened aquatic ecosystems in the world [2]. Because water flow is the driving force of hydrological connectivity in these ecosystems, they are highly sensitive to a variety of hydroclimatic disturbances, affecting both aquatic and terrestrial ecosystems at various spatiotemporal scales. In particular, longitudinal connectivity in river networks is responsible for critical ecological processes, such as the flow of water, nutrients, energy, and aquatic organisms. The pressures induced by humans in relation to water use and flow regulation, dams and hydromorphological alteration, eutrophication and toxic pollution, overfishing and invasive species are widespread and affect river health [3,4]. For instance, chemical pollutants such as pesticides and other industrial waste threaten 50% of Europe's freshwater ecosystems [5]. These chemical pollutants act as chemical barriers and have come, in addition to physical barriers, to modify freshwater biodiversity [6].

Declines in freshwater fishes are the highest worldwide among vertebrates [7,8]. Stream fishes have complex life cycles, including movements between spatially distinct habitats used for different functions [9,10] that condition the viability of populations [11]. In this context, habitat alteration, fragmentation, and connectivity disruption have various consequences on the habitats used by organisms and their movement abilities. In Europe, several authors focusing on the long-term evolution of fish communities have highlighted the decline of migratory species in relation to long-term increases in human pressures [12–15]. The major causes of the extirpation of European diadromous fish species in the twentieth century include the direct and indirect effects of dams [16,17], which prevent access to habitats that species require to complete their life cycles. In addition, a low dissolved oxygen (DO) level, high temperature, or suspended matter content can prevent upstream or downstream movements. In particular, spawning migration is inhibited during hypoxic episodes in the Loire estuary for allis shad (*Alosa alosa*) [18], as well as under unfavorable temperature and DO levels in the Scheldt River for twaite shad (*Alosa fallax fallax*) [19]. As diadromous fish species move long distances between the sea and river networks to complete their life cycle [20], their presence is an indicator of effective longitudinal connectivity in large river systems. In the case of the combined effects of multiple stressors, such as deteriorating water quality, habitat loss, and reduced accessibility, the understanding of the declines in these species is complicated by the required large scope of the study. As they are capable of recolonizing catchments after large-scale disturbances [21], they are ideal indicators of longitudinal connectivity improvement.

Concrete restoration measures require the development of approaches that are able to consider the cumulative impacts of physical and chemical barriers. To the best of our current knowledge, only indices of cumulative impacts of physical barriers have been proposed; for instance, dendritic indices for diadromous species [22] or the length of river habitat affected by barriers [6]. Of the different methods to model river connectivity, the functional approaches that consider species movements in response to spatial heterogeneity are insightful. In this respect, the concept of the "least-cost" path [23] was recently used to quantify how aquatic habitats facilitate or impede fish movements in riverscapes [24] and seascapes [25]. In addition, it is essential to integrate the historical component to better understand long-term fish assemblage changes and species declines in response to human activities [26]. Historical data on physical and chemical barriers are required to model the historical evolution of functional connectivity. Such data, although limited, are relevant to modeling the effect of barriers over time on fish migration routes, colonization fronts, and distribution (see examples in [16,17]).

This study investigates the historical change in the ecological connectivity of the Seine River, from the sea to Paris, since the mid-nineteenth century. This basin has experienced severe declines and extinctions in its diadromous fish community, but it has witnessed the recent recolonization of some species. We propose an approach that combines the impacts of physical and chemical barriers to evaluate the Seine River accessibility from the sea to Paris over time. We applied this approach on three species that differ in their migratory behavior to provide specific guidance to connectivity restoration strategies.

2. Materials and Methods

2.1. Context of the Study Area

The Seine River has a long history of human presence and impacts dating back to the Gallo-Roman era and coinciding with the development of the city of Paris and its activities [27]. The most critical period for the health of the river ecosystem started with the Industrial Revolution in the mid-nineteenth century in France, identified as a turning point with the beginning of the Anthropocene [28]. Deeply modified for navigation and harbor development, the Seine River has undergone various morphological alterations [29], been equipped with navigation weirs, and lost a great number of sandbars, intertidal areas, and islands. Used as a receptacle for liquid waste [28], the river suffered continuous degradation and pollution related to the increase in urban population and industrial activities from the beginning of the nineteenth century [27,30]. In 1889, sewage farms were built upstream and downstream of Paris, but never in a sufficient number to treat all Parisian sewers [30]. After the Second World War, the Achères wastewater treatment plant was established to treat waste from up to eight million people, releasing the treated effluent 70 km downstream of Paris [31]. For nearly five decades, water quality has improved by means of regulation, planning, and management efforts. Currently, the Seine River Basin (76,260 km^2) represents 25–30% of French industrial activity, 50% of national river freight (1400 km of navigable waterways), and 23% of the French population for only 12% of the French territory [31]. The navigation weirs of the Seine River and its main tributaries currently include 23 fish passes.

Before human intervention, the Seine River hosted 11 diadromous and at least 22 freshwater fish species [32]. As in several other European rivers, the fish community has changed since the medieval period due to overfishing, pollution, habitat destruction, and the disruption of migration routes. This led to a dramatic decline in diadromous fishes at the end of the nineteenth century [33]. However, as a result of the establishment of non-native species, diversity has increased and now reaches 60 species [12]. The study area considered for the modeling of historical ecological connectivity is located on the lower Seine River from the sea to Paris and represents around 350 km. The estuary is 160 km long with tidal influence ending at the first obstacle from the sea: the Poses weir (Figure 1). Under the European Water Framework Directive (WFD), the Seine River has been classified as a heavily modified water body, and a recent evaluation (2019) showed contrasting situations from the sea to Paris. Its ecological status is medium to bad in the Seine estuary, medium in the fluvial reach (upstream of Poses) and good from the Oise confluence to Paris. Its physico-chemical status is bad on the whole estuary, medium and poor downstream of the Oise River, and good up to Paris. Mitigation actions are still required to reach "good potential" in 2027, particularly in the context of increasing human pressures on this basin. Notably, effective action is required to strengthen the recovery and sustainability of migratory fish populations in the Seine Basin and are part of numerous regional and national planning documents.

2.2. Selected Fish Species

Among the 11 diadromous fish species historically present in the Seine River Basin, we focused on Atlantic salmon (*Salmo salar*, L., 1758), allis shad (*Alosa alosa*, L., 1758), and sea lamprey (*Petromyzon marinus*, L., 1758), which exhibit contrasting migratory behavior and different patterns of decline. For these three species, which need to migrate upstream of the Seine estuary to find suitable spawning habitats, barriers across the main-stem Seine and its tributaries represent the most significant limiting factor for their current recovery.

Before our study period, a strong decline of Atlantic salmon populations in Northwestern Europe between the early Middle Ages (450–900) and early modern times (1600) was documented and attributed to improvements in watermill technology [14]. In the Seine River, salmon was labeled as a rare species as early as the beginning of the seventeenth century [34] due to overfishing and headwater stream modifications [35]. Today in France, allis shad is critically endangered, the sea lamprey is endangered (a status that has worsened since the last evaluation in 2010), and Atlantic salmon is nearly threatened [36] (Table 1).

All species are listed in the Bern Convention (Appendix V) and in the Habitats Directive of the European Union (Annexes II and V). Migration period(s), DO preferences, and swimming behavior were considered as relevant variables to characterize upstream spawning migration (Table 1).

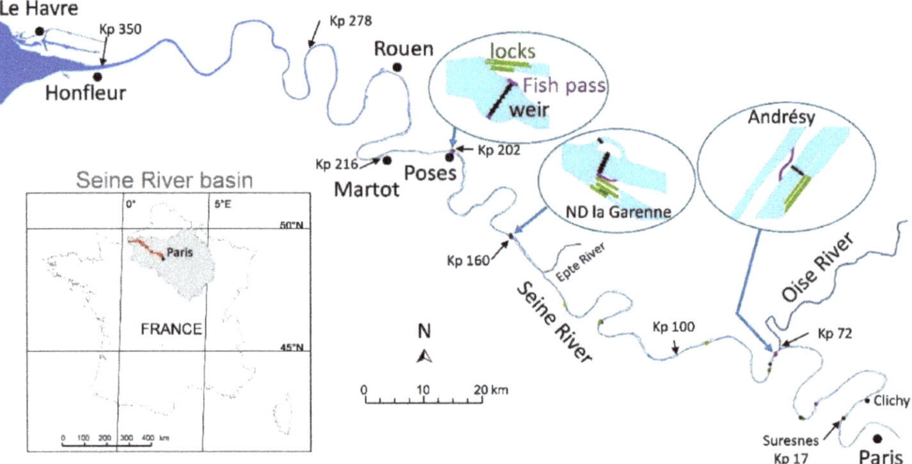

Figure 1. The spatial extent of the studied Seine River from the sea to Paris and its location (in red) in the Seine River Basin in France (thumbnail). The estuary extends up to Poses (kp 202). Some kilometric points are indicated from Paris (kp 0) to Honfleur (kp 350). Zoom-ins on some physical features are illustrated, namely three navigation weirs, locks, and fish passes (2018).

Table 1. Species characteristics for mature migrating individuals. CR: critically endangered; EN: endangered; VU: vulnerable; NT: nearly threatened. Migratory periods were defined using video-counting data (2008–2018) at the Poses fish pass station. Adapted from [37,38], completed: ° [39], * [40]. Atlantic salmon ages are 1SW: one winter at sea, 2SW: two winters at sea.

	Allis Shad	Atlantic Salmon	Sea Lamprey
National status 2019	CR	NT, VU (Allier River population)	EN
Dissolved O_2 $(mg·L^{-1})$	>4	≥6	>3 *
Swimming capacities: burst speed, $(m·s^{-1})$	3.5–5.0	4.5–6.5	3.0–4.5
Behavior to cross obstacle	Swimming	Swimming, high jumping ability	Swimming and "burst and attach" (suction cup mouth)
Migration size [Seine River]	45–70 cm	50–100 cm [1SW < 75 cm, 2SW < 90 cm]	60–90 cm
Migration period(s) (Seine River)	March to June	• March–May • June–July • September–November	March to June

2.3. Historical Data

We defined the main historical periods of physical and chemical changes in the Seine River on the basis of a review of existing literature and expert knowledge. Based on the available data, three periods were chosen as representative of major changes, and these were used for ecological connectivity modeling, namely the 1900s, 1970s, and 2010s.

2.3.1. Physical Features and Infrastructures

We georeferenced the historical maps to compare the river course and the location of physical features across the different periods. For 1900, 89 paper maps of the Seine River from Paris to

Rouen were available thanks to the cartographer Raoul Vuillaume (1:10,000) and from Rouen to the sea thanks to the cartographers Cardin and Babin [41] and the Archives from the Grand Port Maritime de Rouen (GPMR). For 1970, we used the topographic maps of France (1:50,000) produced in 1950 by the National Institute of Geographic and Forest Information (IGN) (available online at https://remonterletemps.ign.fr) and 46 topographic maps (1: 10,000) centered on the Seine River from Paris to Rouen. These high-definition historical images were integrated into a georeferenced system (Lambert93, ESPG 2154) using a thin-plate spline transformation based on standard points (churches, bridges, buildings, crossroads, etc.) and a cubic sampling method that limits geometric distortions. For the current period, we used the National topographic database BD TOPO® 2.2 (IGN, http://www.ign.fr/) where water surfaces are available with two-dimensional (2D) precision from 1.5 to 2.5 m.

Among the physical barriers that could affect fish movements and migrations from the sea to Paris, we focused on weirs, locks, and hydropower plants. Gray literature provided knowledge of the general development of the Seine from Paris to the sea since the nineteenth century [42–45]. The location of weirs and locks, their creation date and modifications, and the evolution of waterfall height were recorded from ancient maps and national and departmental archives (Supplementary Table S1). In particular, the archives of the navigation service of the Seine River contained useful resources such as local maps, longitudinal profiles (with altitudes), detailed maps of weirs and locks, navigation-improvement planning, channel rectification, and island removal. In addition, we used navigation guides, online aerial photos dating back to the First World War (IGN), and online reports from the Bibliothèque Nationale de France (https://gallica.bnf.fr/). All this information was used to evaluate the potential impact of the physical changes on migrations of the three migratory species.

As fish ladders and passes were rarely indicated on historical maps, we consulted other written sources. Thanks to fishermen complaining about the decline of fish stocks and partly accusing navigation dams, the government's awareness of this issue led to decrees and laws, starting with the Fishing Act in 1865. Ministerial circulars about fish passage characteristics, synthesis, and reports from several commissions and surveys about fish ladder effectiveness were used as the main sources to understand their changes. These archives were mainly available in the National Archives and the Archives from the Ministry of Public Works [46]. Knowledge of current fish pass construction is well-documented thanks to reports from the French Navigation Rivers service (VNF) and state services ensuring the regulatory supervision of their effectiveness.

2.3.2. Chemical Barriers

We chose dissolved oxygen (DO) as one of the most relevant factors of water quality affecting fish migration. The information was available for several periods based on different data sources. Monthly/quarterly measurements were available for 22 stations on the Seine River from Paris to Rouen between 1871 and 1938, recorded by the chemical department at the Montsouris Observatory [47,48]. The Water Agency of the Seine-Normandie Basin (created in 1964 to monitor the quality of aquatic systems) provided monthly measures of DO dating back to 1971 for 41 stations from Paris to Honfleur. The Seine navigation service provided monthly measures of DO for the period 1955–2015 from Poses (kp 202) to Honfleur (kp 350). These sources allowed us to create three spatiotemporal databases (kp of available stations × 12 months) of monthly average oxygen values (averaged across around 10-year periods) for each time period, namely 1892–1904, 1971–1980, and 2009–2017. We used linear interpolation to impute missing values using R 3.6.0 [49]. To obtain continuous values through the spatiotemporal table (kp × 12 months), we used the *interp* function from the Akima package (v0.6-2 [50]), which allows for the interpolation of values from irregularly spaced input data [51]. Isopleth graphs were then realized using the *filled.contour* function from the *graphics* package (v3.6.2; [52]). The full reproducible code is available in the Supplementary Materials.

2.3.3. Fish Historical Distribution

To document past species distribution from the end of the eighteenth century to the mid-twentieth century, we used the CHIPS database (Historical Catalog of Fishes of the Seine Basin), which compiles historical written sources [53]. Fish observations from the CHIPS database were updated, georeferenced, and used to map the historical distributions of the three studied species according to the locations of the most upstream presences on the hydrographic network [54]. To map the current distribution, we used recent observations from different sources, such as video-counting at fish passes since 2008 (Seinormigr, personal communication), anglers' catches, and electrofishing surveys. Potential colonization distance estimates were calculated using current hydrographic distances (without considering potential channel modifications) from Honfleur to the most upstream presence recorded in around 1850.

2.4. Functional Connectivity Modeling

The current water extent (vector database 2018, IGN) was manually modified to delineate the water extent for the 1900s and 1970s using the corresponding georeferenced historical maps. The mean DO classes, calculated for each year, species migration period, and kilometer point (kp) were affected to the corresponding reaches using a spatial join procedure in ArcMap. The presence of infrastructures (weir, lock, fish pass, and hydroelectric power plants) was manually digitalized as vector patches using channel extent, historical maps, and available aerial photos. As fish ladders had small widths, their size was enlarged to reach a minimal patch size of 10 × 10 m, which was compatible with the raster modeling resolution of 5 m. All infrastructures were overlaid on 1-km-long reaches with DO classes to generate layers of physical and chemical barriers (hereafter called "resistance maps") for each time and species migration period.

We then ran the least-cost calculation on each resistance map using Anaqualand 2.0 [55]. The minimal cumulative resistance (resistance × distance) or functional distance (functional kilometers: kmf) was calculated for each pixel of the map to obtain accessibility maps from the sea to Paris for each species/migration period and the three time periods. Resistance values were based on the assumption that resistance increases with energy cost, mortality, and predation risk associated with migration. Species-specific resistance values for weirs, locks, fish ladders, and passes were based on their permeability using expert classification according to the different periods studied. To evaluate the relative impact of chemical and physical barriers, we drew accessibility maps that included only physical barriers or maps that included both physical and chemical barriers.

3. Results

3.1. Historical Timeline of Weirs and Locks

When the Amfreville weir (kp 200) was built in 1850 just upstream of Poses, in the naturally tidal part of the Seine, it was the first physical barrier from the sea (Figure 1). To further increase the navigable areas without tidal effect, the Martot weir was built 16 km downstream in 1864 (waterfall height: 0.3 m at high tide −3 m at low tide) followed by the Poses weir in 1881 (no tide influence: 4.18 m [56]). Overall, between 1846 and 1886, the first set of 12 navigation weirs were built from the sea to Paris, of which 10 are on the main channel (Figure 2).

The natural waterline at a low flow in 1840 allowed for a 0.7-m draught; it was transformed in a succession of deep and low-current velocity basins with a draught reaching 3.2 m in 1900. Several technologies were used over time to build navigation weirs, which had increasing impacts on fish passage (Figure 2). The first weirs were composed of thin needles of wood or rolling-up curtain leaned against a solid frame that could be added or removed by hand to constrict river flow (Figures 2 and 3a). Fish could pass through the openings of "needle" weirs during releases for navigation, floods, and winter. The former navigation weirs were renovated in the twentieth century or destroyed to reduce the number of forebays, especially since 1959 due to significant traffic development [57]. Gate dam types (Aubert's, radial, slide gate, and automated systems of flap gates), composed of several gates moving around a horizontal or vertical axis, were built since 1930 (Figure 2). As a result of these technological advances, the number of navigation weirs decreased over time, whereas their waterfall height increased to address navigation needs, impeding more and more fish from passing through, except during extreme floods (Figure 3b). Parallel to this, the activity of narrow locks decreased due to vessel enlargement, leading to some disused open or closed locks. The adaptation of lock operations as means to improve fish passage has not been implemented on the Seine River.

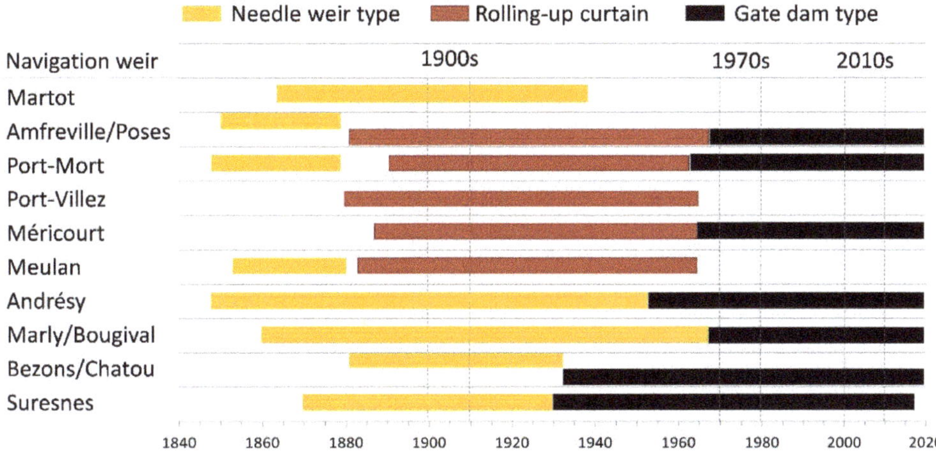

Figure 2. Technology and dates of weirs built on the Seine River from the sea to Paris. The weir at Marly was associated with hydraulic machinery first built in 1684 to raise the water up to an aqueduct tower supplying The Chateau of Versailles. When two names are indicated and bars are vertically staggered, the location of the navigation weir shifted.

3.2. History of Fish Passes

The history of fish passes actually began with the Fishing Act in 1865, which commanded the construction of fish ladders while compensating the dam owners. An evaluation of built fish ladders carried out in 1875 showed the lack of effectiveness of the two fish ladders built on the Seine River: Martot and Marly (Supplementary Table S2). In 1881, using information on the 54 fish ladders existing in France, navigation engineers from the Ponts et Chaussées (bridges and highways) tried to improve ladder efficiency. Main issues were (i) too high slopes, (ii) inadequate current velocities to swimming capacities (target species: mainly Atlantic salmon and shads), or (iii) unattractive locations. As a result, chief engineer Caméré designed several fish ladders from 1890 to 1903 on the Seine River (Supplementary Table S2). Three ladders were specifically designed for eels in the shape of ascending ramps. At this time, only six ladders occurred on the Seine River (Figure 3a) comparing to 100 ladders on the Loire River, and about 10 on the Rhone River). In 1897, a commission on agricultural and forestry improvements concluded that major economic problems would occur in the industrial and agricultural sectors if new fish ladders were prescribed (AN F14-13615). Despite this, decrees ordering French rivers to the fish ladder regime were enacted, especially in 1904 for the Seine River.

A report from the engineer Bachelier put forward a French Atlantic salmon ladder program at the Migratory Fish Commission in 1950 (AN-19920558-18). The scientific laboratory testing of ladder configurations to search for efficient fishways started with Denil's fishway at the beginning of the twentieth century. This was followed by many tests of fishway configurations in different countries [58,59]. An increasing number of new fishways were designed in France (300) in the 1980s following a new decree (1969–1974) in which the construction of fish passages was the responsibility of the owner of the structure. The Fishing Law of 1984 reasserted the classification of watercourses and the obligation to equipped barriers. In 2006, the French law on water and aquatic environments

(LEMA 2006-1772 law), which is the transposal of the European WFD, led, in 2013, to the classification of the Seine River as a watercourse in which it is important to ensure ecological connectivity for mobile organisms and the natural transport of sediments. Its implementation resulted in the acceleration of the renewal of fish passage construction, particularly those equipping the navigation weirs owned by the VNF in the Seine River Basin. Depending on land availability, new passes are of two types: vertical slot pass and bypass channel (Figure 3c).

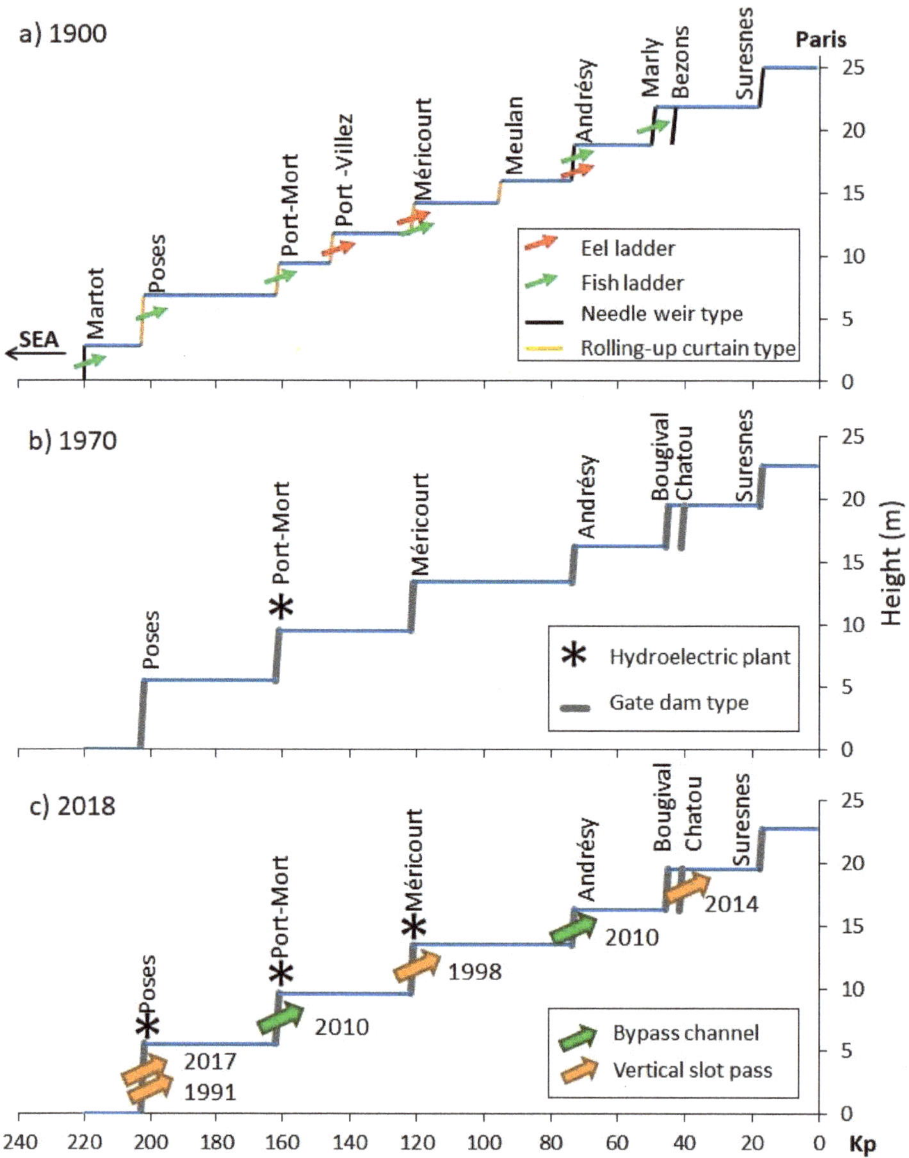

Figure 3. Longitudinal graphs of the location and cumulative fall height of weirs in 1900 (**a**), 1970 (**b**), and 2018 (**c**). Weir types, fish ladders, and passes, as well as hydroelectric plants, are indicated for each time period. The dates of construction of recent fish passes are specified beside arrows.

3.3. Dissolved Oxygen Evolution

Isopleth graph analysis showed strong intra-annual variations of DO from the sea to Paris in the three periods (Figure 4a). The longitudinal patterns appeared to contrast among the three periods and, overall, to be consistent with previous knowledge on the basin [46,60,61]. DO thresholds of 3, 4, and 6 mg·L^{-1} constitute chemical barriers for the sea lamprey, allis shad, and Atlantic salmon, respectively.

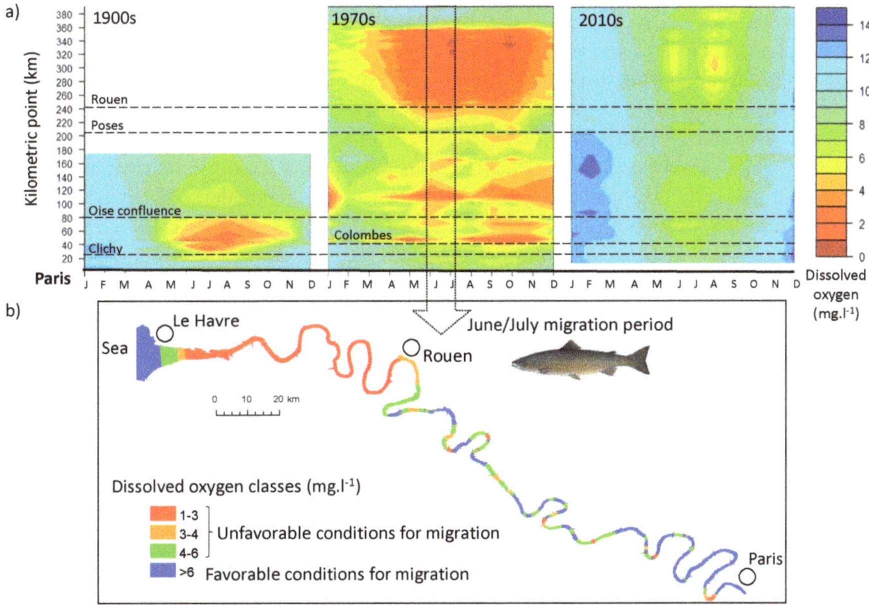

Figure 4. (a) Isopleth graphs of dissolved oxygen for the three time periods. Particular kilometric points are represented, referring to Figure 1. **(b)** An example of the longitudinal mean of dissolved oxygen classes (resolution 1 km) in the 1970s for one migration period of Atlantic salmon (June–July).

In the 1900s, from the Oise confluence to Clichy, unfavorable conditions were observed for the migration of Atlantic salmon, allis shad, and sea lamprey over five-to-seven months. In the 1970s, several hypoxic reaches appeared along the entire Seine River from March to December. A nearly year-long chemical barrier was observed in the estuary from the Seine Bay to Rouen with very low DO values and unfavorable migration conditions for Atlantic salmon in a critical part of its migration route (Figure 4b). Reaches from the Oise confluence to Colombes were impacted by increasing urbanization, diffuse waste, and untreated sewage during rainy events. In the 2010s, no hypoxic periods were observed irrespective of the season, except for a short period in August in the estuary that was not favorable for Atlantic salmon migration.

3.4. Accessibility: Comparison of Functional Distances

Based on historic and current knowledge of physical and chemical barriers, we assigned resistance values according to the potential increasing biological cost/risk associated with crossing the feature (Table 2). The detail of resistance values assignment is provided in Supplementary Material (Online Resource 2).

Accounting only for physical barriers, functional distances to reach Paris from the sea were higher in the 1970s and similar for the three species Figure 5a). In the 1900s, functional distances were lower for Atlantic salmon and allis shad but remained high for sea lamprey due to its lower capacity to cross needle-type weirs and fish ladders Figure 5a). In the 2010s, functional distances were lower due to the fish pass equipment of navigation weirs, but they increased drastically just downstream Paris due to the Suresnes impassable weir. The cumulative impacts of physical and chemical barriers on allis shad and sea lamprey accessibilities to Paris allowed us to examine their different migratory behaviors Figure 5b). In the 1900s, the functional distance increases from the sea to Paris for the two species resulted mainly from the presence of needle and rolling curtain weirs equipped or not with fish ladders. Accessibility for sea lamprey was lower than for allis shad due to its lower capacity to cross the weir and fish ladder (Table 2). Only slight cumulative effects of the chemical barriers were observed between some weirs, as DO values were mainly favorable during the migration period for these two species (March to June) (see Figure 3a). This trend intensified in the 1970s, as weirs became impassable with no fish passes and new anoxic zones cumulated, especially in the estuarine part, affecting allis shad migration. In the 2010s, no chemical barrier was considered (since oxygen was not limiting). Thus, the overall accessibility in this period was similar to the physical barrier-only scenario. Accessibilities were relatively similar for both species with the drastic impact of the Suresnes impassable weir that is downstream from Paris Figure 5b). For Atlantic salmon, we noticed interesting seasonal variations in accessibility in the 1900s and 1970s Figure 5c). The functional distances in the 1900s for the spring and autumn migrations were lower than for the summer migration, indicating better accessibilities for these two seasons. The longitudinal contribution of the scenario with physical barriers on overall functional distances (scenario with physical and chemical barriers) indicated that this stressor dominated for spring migration and its impact decreased in favor of chemical barriers for other migration period (Supplementary Figure S1). In the 1970s, the clear cumulative effects of chemical and physical barriers were observed on accessibilities. In the estuary, functional distances sharply increased due to long-distance chemical barriers, and then a succession of high increases (physical barriers) were cumulated with lower increases (chemical barriers). In this period, summer was still the more unfavorable season for migration with nearly five times lower accessibility than in the 1900s due to hypoxic conditions for long distances along the entire river stretch. Spring was the more propitious season for migration, although conditions were degraded compared with 1900. In the 2010s and currently, migration conditions for this species have been improved compared to the 1900s as the result of fish passes building and decreasing chemical barriers regardless of the season.

Table 2. Values of resistance (dimensionless) for the different type of physical and chemical barriers and associated biological costs and risk. Predation risk is associated with potential predation by other species or poaching when crossing physical barriers. For the accessibility scenario with only physical barriers, the resistance values marked with an asterisk (*) become R = 1.

	Type of Barrier	Biological Cost/Risk	Longitudinal Barriers Thickness (m)	A. Salmon	S. Lamprey	A. Shad
Physical	Hydroelectric dam		10	80,000	80,000	80,000
	Gate dam		10	80,000	80,000	80,000
	Lateral fish ladder		10	80,000	80,000	80,000
	Operating/closed lock	Energetic cost/predation	100	8000	8000	8000
	Needle weir		10	40,000	60,000	40,000
	Rolling curtain weir		10	40,000	60,000	40,000
	Fish ladder		10	10,000	20,000	10,000
	Fish pass: vertical slot		20	1000	2000	1000
	Fish pass: secondary channel		200	50	200	150
	Disused open lock	Minimal cost	100	2	2	2
Chemical	Reach oxygen class 1–3 mg·L^{-1}	Mortality/physiologic cost	1000	100 *	10 *	20 *
	Reach oxygen class 3–4 mg·L^{-1}		1000	20 *	1 *	10 *
	Reach oxygen class 4–6 mg·L^{-1}	Minimal cost	1000	10 *	1	1
	Reach oxygen class > 6 mg·L^{-1}	No cost assumed	1000	1	1	1

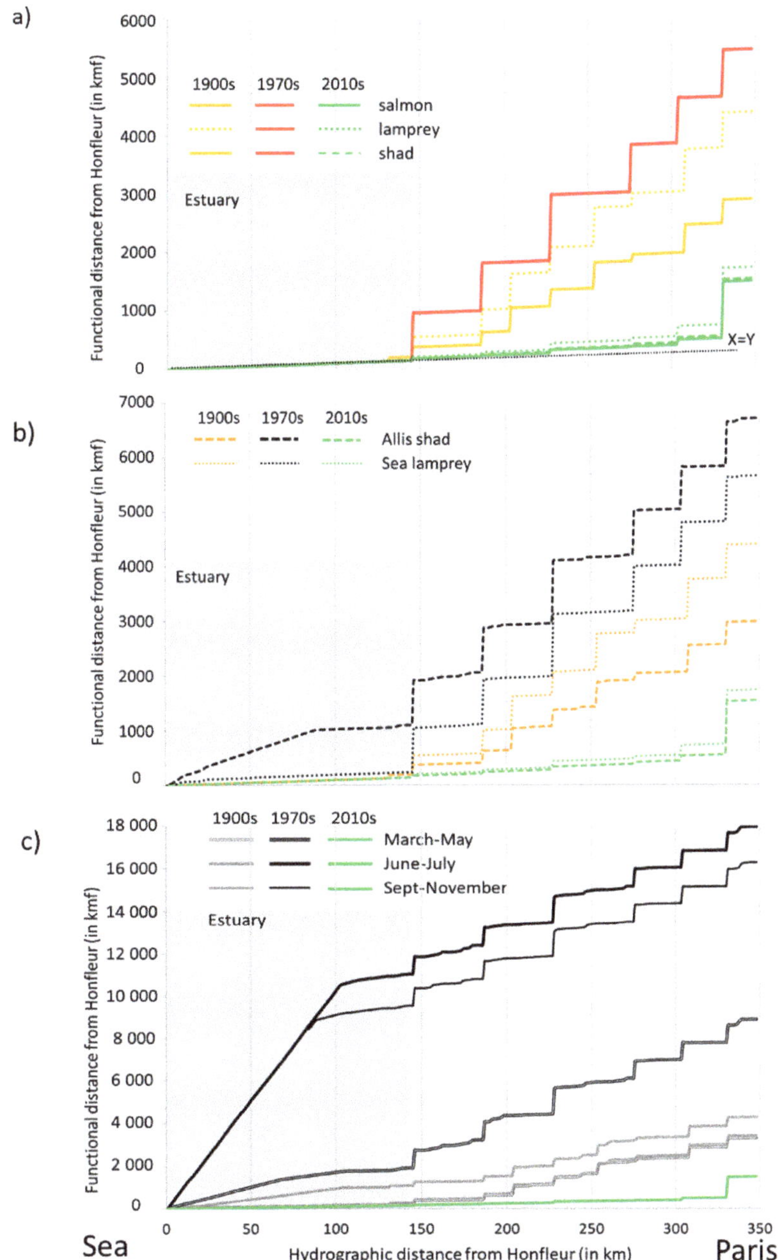

Figure 5. Accessibility from the sea to Paris, calculated in functional kilometers in relation to hydrographic distances. (**a**) Accessibility calculated with only physical barriers for the three species and time periods. The X = Y line is indicated. (**b**) Accessibility calculated for allis shad and sea lamprey for the physical and chemical barriers scenario and time periods. (**c**) Accessibility for Atlantic salmon in relation to migration periods and time periods.

3.5. Historical Fish Distribution

In the 1850s, Atlantic salmon was known to reach headwaters of the basin to spawn up to 684 km from the sea on the Cure River, but occasional observations were also recorded on the Marne (610 km) and the Aisne (528 km) Rivers in the northern part of the basin (Figure 6a).

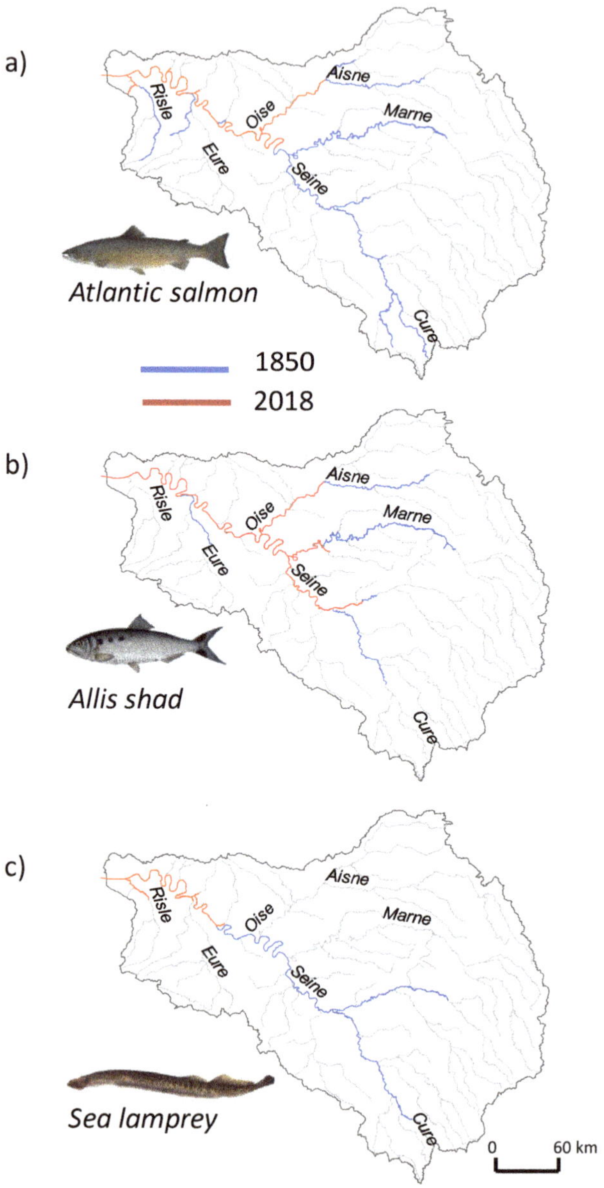

Figure 6. Historical distribution in 1850 overlaid by current distribution in 2018 for Atlantic salmon (**a**), allis shad (**b**) and sea lamprey (**c**) on the Seine River Basin. No distribution was drawn for 1970, as all these three species were regarded as extirpated at that time.

The second half of the nineteenth century showed the continued collapse of Atlantic salmon, then restricted to only some tributaries, due to multiple barriers (weirs and locks, dams in headwater catchments, and pollution) to fish spawning migration [39]. In the 1970s, this species was regarded as extirpated [12], despite individuals caught accidentally in the lower part of the estuary (kp 278) [62]. In 2008, 260 individuals of Atlantic salmon were observed passing at the video-counting station of Poses [37], but there was no indication of their distribution upstream. The known current distribution extends up to 350–400 km to the Oise River (Figure 6a) as indicated by records in a new video-counting station at the confluence of the Aisne River (2018). Allis shad was one of the most widely spread migratory species in the Seine Basin in the first part of the nineteenth century [40]. The most upstream reports of its presence were 624 km away from the sea on the Marne River, 538 km on the Aisne River, and 551 km on the Yonne River (Figure 6b). Its decline started with the first navigation weirs built between 1850 and 1881. In 1920, only rare and isolated individuals were still observed downstream of Poses weirs [41]. Allis shad was considered to be extinct in the 1960s [42]. More broadly, the decline of shads has been related to the cumulative effect of habitat degradation and physical and chemical barriers preventing individuals from reaching spawning habitats upstream of estuaries [43]. In the Seine River, recolonizing individuals of allis shad have been observed since 2004, with clear evidence of reproduction [40], and, today, they have been observed up to 400–454 km in some tributaries (Figure 6b). In the 1850s, sea lamprey had two attested colonization fronts in the Cure River Basin (614 km) and the Aube River (586 km), but we can speculate that it had a wider distribution (Figure 6c). The freshwater distribution of the sea lamprey decreased over the second half of the twentieth century, and the species was considered extinct in the 1970s [42]. The current distribution of sea lamprey extends up to 230 km (Figure 6c).

4. Discussion

4.1. Reconstructing the Tide of History

By combining historical data sources, engineering sciences, and fish ecology, this interdisciplinary study offers a better understanding of how multiple stressors act on diadromous fish species in the Seine River over a long-term perspective. The use of historical sources and current data on physical and chemical barriers made it possible to reconstruct the history of ecological connectivity for fish from the sea to Paris over the past 170 years. Precise knowledge of the chronological timeline of the two stressors provided a sort of spatiotemporal puzzle by using the history of each piece acting positively or negatively on ecological connectivity. Such socio-ecological hydrosystems are complex adaptative systems that have unexpected emergent properties that cannot be predicted by knowing the individual constituents alone [63]. While global trends in the spatial distribution of physical and chemical stressors have been documented in the Seine River [64], we have quantified their cumulative effects on fish migration over time. Both stressors have been documented in other large river systems in a long-term perspective, but they have often been considered separately [16,65–67].

The temporal trajectories of species decline and recolonization, visible over the spatial extent of the species distribution over time, have been related to the ecological connectivity changes. The migratory fish species in the Seine River Basin have generally followed the common pattern of declines observed for North Atlantic diadromous species, namely a sharp population decline between the end of the eighteenth and the beginning of the twentieth century [68]. The distribution of the three studied species in the 1850s extended in the Seine River and main tributaries up to 500–700 km upstream from the sea, although this situation could have already been the result of early declines since the Middle Ages [14]. During the 1850–1881 period, the most downstream Martot and Amfreville/Poses weirs (cumulative height of 6.8 m over just 20 km) are known to have had a major impact by reducing the accessibility of a wide part of the basin [69]. The construction of the first 12 navigation weirs (1846–1869), the delay of most fish ladder construction (1880–1903), and their poor effectiveness led to the collapse of Atlantic salmon stocks in the 1900s and to the extirpation of allis shad in the 1920s [33].

Because of the increasing impact of physical barriers, most migratory species also became extinct at this period in the Lambro River (Milan), the Spree River (Berlin) [70], and the Scheldt River [71]. In the 1900s, we highlighted that chemical barriers were already present in the fluvial and estuarine reaches of the Seine River. The deficit in DO around 70–150 km downstream of the sewer system discharge in Clichy was compensated for by flows from the Oise confluence, which played a crucial role in reoxygenating the Seine [60]. The seasonal differences in accessibilities to Paris estimated in the 1900s could have particularly affected young salmons (one winter at sea (1SW)), since they migrate from February to September, compared with the oldest salmons (two winters at sea (2SW)), for which migration started in autumn [46]. However, historical sources have suggested that the Seine Atlantic salmon stock was composed mainly of large individuals (2SW). Parallel to this, historical Atlantic salmon spawning grounds on the Cure River (see its location Figure 6) became unreachable in 1858 with the construction of the Settons Dam (19 m high) [72].

The highest cumulative disruption of ecological connectivity was observed in the 1970s as a consequence of the post-war boom with a period of strong industrialization. The long-distance chemical barrier in the Seine estuary (and many more along the river) concurred with the renovation and heightening of weirs with no fish passes, thereby explaining this result. The very low level of DO in the Seine estuary was principally related to inputs from its upstream watershed [73]. In the 1970s, more than half of the wastewater produced by Paris was discharged into the Seine without treatment. Tributaries such as the Oise River were no longer playing the role of re-oxygenating the Seine River. The alteration of water quality was such that the extinction of species was considered irreversible and the idea of maintaining and rebuilding fish passes was abandoned. Integrating all these issues, the highest functional distances from the sea to Paris were observed at this period, especially for Atlantic salmon in summer and autumn migrations, as well as, to a lesser extent, for allis shad and sea lamprey. Whereas the first two species were still considered as extirpated from the Seine River Basin, the sea lamprey disappeared in the 1970s.

The current ecological connectivity has improved at a highest level than in the 1900s for the three species. In the 2010s, favorable oxygenating conditions were observed for the three species, and all migration periods in parallel to the construction of a new generation of efficient fish passes. Our results confirm the recent trend of no further long anoxic periods in the Seine River and its estuary since 2007 [61], a consequence of the progress made in the 1990s in terms of wastewater treatment following the Water Law (1964). At the same time, the Fishing Law of 1984 revived the construction of fish passages. The "return to sources contracts," which proposed the first management plans for migratory fish, was established, and several migratory fish associations in the French river basins were created. In this context, a study defined the strategy for the return of Atlantic salmon to the Seine River at the beginning of the 1990s [74], and the first fish pass of Poses was built in 1991. The EU WFD and the National Plan for ecological continuity (2010) just reinforced this turning point, and the renewal of fish passage construction was deployed during the recent renovation of navigation weirs. The recent improvement in migration route accessibility has very recently led to the spontaneous recolonization of the Seine River by individuals of migratory fish species. Since 2004, some individuals of allis shad have been observed upstream of Paris on the Marne and Seine Rivers. Their surprising presence upstream of an navigation weir with no fish pass was probably related to high flows and/or the ability of this species to use locks, as shown with radio-tracking operations in the Rhone River [75]. On the Oise River, where all the navigation weirs are equipped with fish passes, the most upstream observations of Atlantic salmon and allis shad were reported in 2018 by video-counting at the Aisne confluence (Figure 6). To our knowledge, the distribution of sea lamprey remained restricted up to the Epte River, where the local angling association observed reproduction events (see Figures 1 and 6). However, the observation of one individual on the lower Oise River in the early 2000s (Holl, personal communication) could suggest further potential for settlement.

The global historical analysis highlighted that, in addition to structural aspects, there is social aspect in explaining the changes in ecological connectivity through time. Such river–society interactions are spatially and temporally complex and can be addressed by an interdisciplinary collaboration between hydrobiologists and environmental historians [60]. Our study showed the example of a river long affected by humans, such as the Moselle [76] or the Danube [67], as a very complex historical object. We emphasized the interactions of human impacts on functional connectivity with historic conflicts of use between stakeholders regarding the attempt to restore the free movement of migratory fish. The social consensus, driven by industrial interests for long periods [76] and the perception of rivers as not healthy, has, for instance, prevented the implementation of efficient fish passage management strategies. Beyond the Seine, lessons from our historical approach to ecological connectivity can be applied to other riverine anthroposystems affected by physical and chemical barriers. In particular, since many large river systems in Northwestern Europe have been affected by common nineteenth and twentieth century stressors such as river regulation and dam construction, pollution, and overfishing [77], our approach can be useful for a better understanding of the functional connectivity of these systems.

4.2. The Modeling Approach

The least-cost modeling approach and the functional distance calculation that spatially integrates the cumulative effect of physical and chemical barriers provide a fish-based longitudinal indicator of the habitat accessibility for large rivers. This approach was useful to evaluate the relative contribution of physical and chemical barriers in habitat accessibility and to help disentangle the complexity of multiple-stressor situations. For instance, the comparison of scenarios with and without chemical barriers indicated a longitudinal contribution of chemical stress in the overall accessibility to upper Seine. The spatially explicit approach was also beneficial to visually observe the impact of the two types of barriers along the river course and to compare the access to a specific kilometric point according to the time period for the different species. Maps produced using our approach represent a valuable medium for communicating to managers the consequences of different management scenarios and, subsequently, guiding decision-making.

The combination of historical and current sources (maps, reports, questionnaires, postcards, pictures, etc.) was relevant in order to map and precisely locate navigation weirs, locks, and fish ladders/passes, thereby allowing for longitudinal cumulative calculations of accessibility. The use of functional kilometers compared with hydrographic kilometers was intuitive and made it possible to understand the relative cost of crossing physical and chemical barriers. The comparison between historical periods for behaviorally distinct species highlighted the complex impacts of physical and chemical barriers on their migration capacities and their dispersion to upstream habitats.

The resolution of 1 km we used to spatialize DO values from the sea to Paris was a compromise, as the original data had different spatial resolutions depending on the time period. We imputed missing data for the 1900s with the "best hypothesis and knowledge of historians." Despite this potential bias, we obtained consistent longitudinal DO profiles, in particular for the long-term summer hypoxia that is well documented [31]. Though historical trends in navigation weir settlement are relatively known [31], information such as the precise date of construction, renovation, or removal was difficult to recover or was barely available. Reports from navigation engineers were particularly valuable to document the technological evolution of weirs and fish ladders and to understand the evolution of passability for fish.

While this approach has demonstrated its usefulness in comparing scenarios [78] and emphasizing the relative importance of chemical and physical barriers on connectivity, there is still room for improvement, including refining the hypothesis underlying the choice of resistance values [79]. One perspective is to integrate the individual resistance for each weir based on its height and for each fish pass in the 2010s based on their functioning and expert evaluation. The movement of individuals using acoustic telemetry data could also provide invaluable knowledge about movement behavior, as shown downstream of Poses for estuarine species [80], which could help the calibration of resistance values. Future experiments using acoustic telemetry would provide a more realistic range of resistance values for each recent fish pass (see, e.g., the Consacre project, www.consacre.fr). Another avenue of research is to integrate the flood tide in connectivity modeling, as this process can facilitate species migration upstream from the estuary. In addition, water temperature during migration could be integrated into future scenarios as a potential future chemical barrier, since this environmental parameter, particularly in summer, impacts DO values and could lead to mortality, as shown in the Allier River [81].

4.3. Management Implications and Perspectives

Adaptative management for highly anthropized hydrosystems could benefit from the insights gained from past experience and the knowledge of potential long-term legacies [82]. These elements are fundamental for developing strategies to envisage the mitigation of interacting stressors [83]. Understanding the structural and social interacting causes of the past decline and recolonization processes of fish species guides future management and helps prioritize actions along the Seine River. We have highlighted the dominant role of chemical quality over other physical features as the consequence of historical management actions undertaken from the mid-nineteenth century onwards. The very limited evolution of the situation and regulation of fish ladders and passes since the end of the nineteenth century has resulted from long-run conflicts of interest between fishermen, industry, and agriculture, as well as from the complex role of the Ministry of Public Works in the application of decrees and laws [84]. A 1902 review by Violette with contemporary resonance underlined the prejudice relating to the higher degree of public utility of installing a hydro-electric plant compared with that of the right to fish [85]. Actions in favor of fish passage finally became more easily defensible with the implementation of successive laws —Water Law in 1964 and Fishing Law in 1984— and European regulations in the twentieth century and early twenty-first century. The context of improvement of water quality in the 1980s contributed to opening mindsets to environmental measures. Today, the observation of individuals of Atlantic salmon and allis shad on the Oise River is an encouraging sign, showing the effectiveness of ecological connectivity restoration methods, such as equipping all navigation weirs with a fish pass. In this context, it is crucial to develop complementary tools to measure ecological connectivity, e.g., video-counting or biotelemetry data, as well as the participation of anglers by communicating their catches.

In order to support the raising recovery of migratory fish, there are several alternatives as management actions. First, our study confirmed the importance of maintaining good chemical water quality throughout the year on the Seine River, its estuary, and its main tributaries. In particular, avoiding hypoxic events in the tidal freshwater part of the estuary is important to allow for upstream migration opportunities for anadromous spawners, as also pointed out in the Scheldt study [71]. Improvements in wastewater networks and their management have led to a considerable reduction in discharges during rainy weather, which decreases the risk of seasonal fish mortality. This is crucial in large river systems that are structurally sensitive to wastewater inputs due to low flows or extreme events (violent storms or industrial accidents), which can lead to rapid decreases in the oxygen level below fish survival values [86]. Second, maintaining (and increasing) the effectiveness of existing fish passages is also an important driver and a condition for sustainable restoration measures [87]. Third, the preservation or restoration of ecological connectivity in tributaries close to the estuarine area is another way to improve the maintenance of migratory populations. Further research is needed to identify the locations of potential suitable spawning habitats in the basin (as potential targets of migration). Though the conditions for upstream migration have generally improved over the last few decades, the potential local deterioration or destruction of spawning habitats needs to be evaluated because it strongly limits the possibilities for restoration.

An important issue for the sustainable management of river basins in Europe is to integrate future scenarios of global change. Climate change will have significant effects on the Seine River Basin, including modifying its flood regime [88,89]. This will, in turn, affect the future ecological connectivity for species that are now recolonizing the Seine River Basin. Therefore, modeling these effects is of great importance to guide future management actions. Projections of species distribution over Europe and scenarios of temperature evolution in the Seine River Basin suggest the potential favorability of the basin for shads but a decreasing favorability for salmonids [37]. In this context, prioritizing efforts to restore ecological connectivity could also consist of focusing on cooler tributaries and upstream parts of the Seine, Oise, and Marne Rivers.

Supplementary Materials: The following are available online at http://www.mdpi.com/2073-4441/12/5/1352/s1, Online resource: Resistance values assignment, Figure S1: Longitudinal changes in the contribution of the scenario with physical barriers on the overall functional distances calculated for Atlantic salmon in the 1900s according to the migration periods, Table S1: On-site and online historical and current sources used to document the studied periods, Table S2: Fish ladders built on the Seine River at the end of the nineteenth century. Specific ladders for eel passage are indicated in italic.

Author Contributions: Conceptualization and methodology, C.L.P., L.L., J.B., and E.T.; archives investigation and resources, L.L., E.C., J.B., and M.-L.M.; formal analysis, E.C., C.L.P., and M.-L.M.; writing—original draft preparation, C.L.P. and E.C.; writing—review and editing, C.L.P., L.L., J.B., E.T., and M.-L.M. All authors have read and agreed on the published version of the manuscript.

Funding: This research was funded by the Water Agency "Agence de l'Eau Seine-Normandie" and the Regions Normandie and Ile-de-France.

Acknowledgments: This study was carried out as part of the CONSACRE project, managed by the GIP Seine-Aval and INRAE-HYCAR and supported by the state-interregional planning contract "Vallée de la Seine" (2015–2020) and the ZA Seine (LTSER, Long-Term Ecosystem Research in Europe). We thank Nicolas Bacq, Eric L'Ebrellec, and Cédric Fisson (GIP Seine-Aval) for providing historical documents and georeferenced data of the Seine estuary. We thank Amandine Zahm (INRAE) for running part of the tested scenarios and Olivier Delaigue (INRAE) for his help in writing and running R scripts.

Conflicts of Interest: The authors declare no conflict of interest.

References

1. Thorp, J.H.; Thoms, M.C.; Delong, M.D. The riverine ecosystem synthesis: Biocomplexity in river networks across space and time. *River Res. Appl.* **2006**, *22*, 123–147. [CrossRef]

2. WWF. *Living Planet Report-2018: Aiming Higher*; Grooten, M., Almond, R.E.A., Eds.; WWF International: Gland, Switzerland, 2018; 148p.

3. Dudgeon, D.; Arthington, A.H.; Gessner, M.O.; Kawabata, Z.-I.; Knowler, D.J.; Lévêque, C.; Naiman, R.J.; Prieur-Richard, A.-H.; Soto, D.; Stiassny, M.L.J.; et al. Freshwater biodiversity: Importance, threats, status and conservation challenges. *Biol. Rev.* **2006**, *81*, 163–182. [CrossRef] [PubMed]

4. Dynesius, M.; Nilsson, C. Fragmentation and flow regulation of river systems in the northern third of the world. *Science* **1994**, *266*, 753–762. [CrossRef] [PubMed]

5. Malaj, E.; Von der Ohe, P.C.; Grote, M.; Kühne, R.; Mondy, C.P.; Usseglio-Polatera, P.; Brack, W.; Schäfer, R.B. Organic chemicals jeopardize the health of freshwater ecosystems on the continental scale. *Proc. Natl. Acad. Sci. USA* **2014**, *111*, 9549–9554. [CrossRef]

6. Fuller, M.R.; Doyle, M.W.; Strayer, D.L. Causes and consequences of habitat fragmentation in river networks. *Ann. N. Y. Acad. Sci.* **2015**, *1355*, 31–51. [CrossRef]

7. Vörösmarty, C.J.; McIntyre, P.B.; Gessner, M.O.; Dudgeon, D.; Prusevich, A.; Green, P.; Glidden, S.; Bunn, S.E.; Sullivan, C.A.; Liermann, C.R. Global threats to human water security and river biodiversity. *Nature* **2010**, *467*, 555. [CrossRef]

8. Duncan, J.R.; Lockwood, J.L. Extinction in a field of bullets: A search for causes in the decline of the world's freshwater fishes. *Biol. Conserv.* **2001**, *102*, 97–105. [CrossRef]

9. Fullerton, A.H.; Burnett, K.M.; Steel, E.A.; Flitcroft, R.L.; Pess, G.R.; Feist, B.E.; Torgersen, C.E.; Miller, D.J.; Sanderson, B.L. Hydrological connectivity for riverine fish: Measurement challenges and research opportunities. *Freshw. Biol.* **2010**, *55*, 2215–2237. [CrossRef]

10. Pringle, C.M. Hydrologic connectivity and the management of biological reserves: A global perspective. *Ecol. Appl.* **2001**, *11*, 981–998. [CrossRef]

11. Falke, J.A.; Dunham, J.B.; Jordan, C.E.; McNyset, K.M.; Reeves, G.H. Spatial ecological processes and local factors predict the distribution and abundance of spawning by Steelhead (*Oncorhynchus mykiss*) across a complex riverscape. *PLoS ONE* **2013**, *8*, e79232. [CrossRef]

12. Belliard, J.; Beslagic, S.; Delaigue, O.; Tales, E. Reconstructing long-term trajectories of fish assemblages using historical data: The Seine River basin (France) during the last two centuries. *Environ. Sci. Pollut. Res.* **2018**, *25*, 23430–23450. [CrossRef] [PubMed]

13. Haidvogl, G.; Hoffmann, R.; Pont, D. Historical ecology of riverine fish in Europe. *Aquat. Sci.* **2015**, *77*, 315–324. [CrossRef] [PubMed]

14. Lenders, H.; Chamuleau, T.; Hendriks, A.; Lauwerier, R.; Leuven, R.; Verberk, W. Historical rise of waterpower initiated the collapse of salmon stocks. *Sci. Rep.* **2016**, *6*, 29269. [CrossRef] [PubMed]

15. Lobón-Cerviá, J.; Elvira, B.; Rincón, P. Historical Changes in the Fish Fauna of the River Duero Basin. In *Historical Change of Large Alluvial Rivers; Western Europe*; Petts, G.E., Möller, H., Roux, A.L., Eds.; John Wiley and Sons: Chichester, UK, 1989; pp. 221–232.

16. Segurado, P.; Branco, P.; Avelar, A.P.; Ferreira, M.T. Historical changes in the functional connectivity of rivers based on spatial network analysis and the past occurrences of diadromous species in Portugal. *Aquat. Sci.* **2015**, *77*, 427–440. [CrossRef]

17. Van Puijenbroek, P.J.T.M.; Buijse, A.D.; Kraak, M.H.S.; Verdonschot, P.F.M. Species and river specific effects of river fragmentation on European anadromous fish species. *River Res. Appl.* **2019**, *35*, 68–77. [CrossRef]

18. Tétard, S.; Feunteun, E.; Bultel, E.; Gadais, R.; Bégout, M.-L.; Trancart, T.; Lasne, E. Poor oxic conditions in a large estuary reduce connectivity from marine to freshwater habitats of a diadromous fish. *Estuar. Coast. Shelf Sci.* **2016**, *169*, 216–226. [CrossRef]

19. Maes, J.; Stevens, M.; Breine, J. Poor water quality constrains the distribution and movements of twaite shad *Alosa fallax fallax* (Lacépède, 1803) in the watershed of river Scheldt. In *Fish and Diadromy in Europe (Ecology, Management, Conservation)*; Dufour, S., Prévost, E., Rochard, E., Williot, P., Eds.; Developments in Hydrobiology; Springer: Dordrecht, The Netherlands, 2008; Volume 200, pp. 129–143.

20. McDowall, R.M. *Diadromy in Fishes: Migrations between Freshwater and Marine Environments*; Croom Helm: London, UK, 1988; 250p.

21. Dufour, S.; Prévost, E.; Rochard, E.; Williot, P. *Fish and Diadromy in Europe (Ecology, Management, Conservation)*; Developments in Hydrobiology; Springer: Dordrecht, The Netherlands, 2008; Volume 200, 186p.

22. Cote, D.; Kehler, D.; Bourne, C.; Wiersma, Y. A new measure of longitudinal connectivity for stream networks. *Landsc. Ecol.* **2009**, *24*, 101–113. [CrossRef]

23. Adriaensen, F.; Chardon, J.P.; De Blust, G.; Swinnen, E.; Villalba, S.; Gulinck, H.; Matthysen, E. The application of 'least-cost' modelling as a functional landscape model. *Landsc. Urban Plan.* **2003**, *64*, 233–247. [CrossRef]

24. Foubert, A.; Le Pichon, C.; Mingelbier, M.; Farrell, J.M.; Morin, J.; Lecomte, F. Modeling the effective spawning and nursery habitats of northern pike within a large spatiotemporally variable river landscape (St. Lawrence River, Canada). *Limnol. Oceanogr.* **2019**, *64*, 803–819. [CrossRef]

25. Caldwell, I.; Gergel, S. Thresholds in seascape connectivity: Influence of mobility, habitat distribution, and current strength on fish movement. *Landsc. Ecol.* **2013**, *28*, 1937–1948. [CrossRef]

26. Shaffer, H.B.; Fisher, R.N.; Davidson, C. The role of natural history collections in documenting species declines. *Trends Ecol. Evol.* **1998**, *13*, 27–30. [CrossRef]

27. Guillerme, A. Le testament de la Seine/The legacy of the Seine. *Géocarrefour* **1990**, *65*, 240–250. [CrossRef]

28. Meybeck, M.; Lestel, L.; Briand, C. La Seine sous Surveillance: Les Analyses des Impacts de L'agglomération Parisienne par L'observatoire de Montsouris de 1876 à 1937. In *Les Rivières Urbaines et Leur Pollution*; Lestel, L., Carré, C., Eds.; Collection Indisciplines, Editions Quae: Versailles, France, 2017; pp. 32–43.

29. Lafite, R.; Romaña, L.-A. A man-altered macrotidal estuary: The Seine estuary (France): Introduction to the special issue. *Estuaries Coasts* **2001**, *24*, 939. [CrossRef]

30. Barles, S. Urban metabolism and river systems: An historical perspective–Paris and the Seine, 1790–1970. *Hydrol. Earth Syst. Sci.* **2007**, *4*, 1845–1878. [CrossRef]

31. Meybeck, M.; Lestel, L.; Carré, C.; Bouleau, G.; Garnier, J.; Mouchel, J.M. Trajectories of river chemical quality issues over the Longue Durée: The Seine River (1900s–2010). *Environ. Sci. Pollut. Res.* **2018**, *25*, 23468–23484. [CrossRef] [PubMed]

32. Belliard, J.; Boet, P.; Allardi, J. Évolution à long terme du peuplement piscicole du bassin de la Seine. *Bull. Fr. Pêche Piscic.* **1995**, 83–91. [CrossRef]

33. Belliard, J.; Beslagic, S.; Tales, E. Changes in Fish Communities of the Seine Basin over a Long-Term Perspective. In *The Seine River Basin*; Flipo, N., Labadie, P., Lestel, L., Eds.; Handbook of Environmental Chemistry; Springer: Cham, Switzerland, 2020.

34. Cavailler, P. Le fief de la Mothe ou maîtrise de l'eau de Corbeil au début du XVIIe siècle. In *Paris et Ile-de-France, Mémoires, t. 45, La Seine et son Histoire en Ile-de-France*; Fédération des sociétés historiques et archéologiques de Paris et Ile-de-France: Paris, France, 1994; pp. 169–192.

35. Benoit, P.; Loridant, F.; Matteoni, O. Pêche et Pisciculture en eau douce: La Rivière et L'étang au Moyen Age. In *Actes des 1es Rencontres Internationales de Liessies (conference proceedings)*; Conseil général du Nord: Lille, France, 2004; pp. 4–5.

36. UICN; MNHN; AFB. *Liste Rouge des Espèces Menacées en France, les Poissons D'eau Douce de France Métropolitaine*; UICN: Montreuil, France, 2019; 16p.

37. Rochard, E.; Pellegrini, P.; Marchal, J.; Béguer, M.; Ombredane, D.; Lassalle, G.; Menvielle, E.; Baglinière, J. Identification of diadromous fish species on which to focus river restoration: An example using an eco-anthropological approach (the Seine basin, France). In Proceedings of the Challenges for Diadromous Fishes in a Dynamic Global Environment, Halifax, NS, Canada, 18–21 June 2007; pp. 691–711.

38. Belliard, J.; Marchal, J.; Ditche, J.-M.; Tales, E.; Sabatié, R.; Baglinière, J.-L. Return of adult anadromous allis shad (*Alosa alosa* L.) in the river Seine, France: A sign of river recovery? *River Res. Appl.* **2009**, *25*, 788–794. [CrossRef]

39. Baudoin, J.; Burgun, V.; Chanseau, M.; Larinier, M.; Ovidio, M.; Sremski, W.; Steinbach, P.; Voegtle, B. Evaluer le franchissement des obstacles par les poissons. In *Principes et Méthodes*; Onema: Paris, France, 2014.

40. Claridge, P.; Potter, I. Oxygen consumption, ventilatory frequency and heart rate of lampreys (*Lampetra fluviatilis*) during their spawning run. *J. Exp. Biol.* **1975**, *63*, 193–206.

41. Cardin; Babin. *Carte de L'estuaire de la Seine Indiquant les Variations du Chenal de 1880 à 1900*; 1907.

42. AESN. *Tome 2: Besoins et Utilisation d'eau, Pollution. Fascicule 6—Hydraulique Fluviale et Voies Navigables*; Les Bassins de la Seine et des cours d'eau Normands: Paris, France, 1976; 140p.

43. *Agence Financière de Bassin Seine Normandie*; Notre fleuve: La Seine, France, 1980; 32p.

44. Lescure, S.; Arnaud-Fassetta, G.; Cordier, S. Sur quelques modifications hydromorphologiques dans le Val de Seine (Bassin parisien, France) depuis 1830: Quelle part accorder aux facteurs hydrologiques et anthropiques? *EchoGéo* **2011**, *8*, 1–18. [CrossRef]

45. Leynaud, G.; Trocheris, F.; Larinier, M. Les obstacles à la réalisation du cycle vital des poissons. *La Houille Blanche* **1987**, 39–44. [CrossRef]

46. Beslagic, S. Histoire des Interactions Entre les Sociétés Humaines et le Milieu Aquatique Durant L'anthropocene. Évolutions des Peuplements Piscicoles dans le Bassin de la Seine. Ph.D. Thesis, Université Paris 1-Panthéon-Sorbonne, Paris, France, 2013.

47. Carbonaro-Lestel, L.; Meybeck, M. La mesure de la qualité chimique de l'eau, 1850–1970. *La Houille Blanche* **2009**, *3*, 25–30. [CrossRef]

48. Meybeck, M.; Lestel, L.; Briand, C. L'impact de l'agglomération parisienne sur le milieu aquatique de 1876 à 1937, dans les travaux de l'Observatoire de Montsouris. In *Les Rivères Urbaines et Leur Pollution*; Lestel, L., Carré, C., Eds.; Collection Indisciplines, Editions Quae: Versailles, France, 2017; pp. 32–42.

49. R Core Team. *R: A Language and Environment for Statistical Computing*; Foundation for Statistical Computing: Vienna, Austria, 2019.

50. Akima, H.; Gebhardt, A.; Petzold, T.; Maechler, M. Akima: Interpolation of Irregularly and Regularly Spaced Data. R Package Version 0.6-2. 2016. Available online: https://www.google.com.hk/url?sa=t&rct=j&q=&esrc=s&source=web&cd=1&ved=2ahUKEwiZ0om8uajpAhUPQd4KHXcKAawQFjAAegQIBBAB&url=https%3A%2F%2Fcran.r-project.org%2Fweb%2Fpackages%2Fakima%2Fakima.pdf&usg=AOvVaw2l4Omur7FN7TZ3F_f2hbUk (accessed on 10 May 2020).

51. Akima, H. A method of bivariate interpolation and smooth surface fitting for irregularly distributed data points. *ACM Trans. Math. Software (TOMS)* **1978**, *4*, 148–159. [CrossRef]

52. Cleveland, W.S. *Visualizing Data*; Hobart Press: Summit, NJ, USA, 1993; 360p.

53. Beslagic, S.; Marinval, M.-C.; Belliard, J. CHIPS: A database of historic fish distribution in the Seine River basin (France). *Cybium* **2013**, *37*, 75–93.

54. Merg, M.L.; Dézerald, O.; Kreutzenberger, K.; Demski, S.; Reyjol, Y.; Belliard, J. Modeling diadromous fish loss from historical data: Identification of anthropogenic drivers and testing of mitigation scenarios. *PLoS ONE*. (under review).

55. Le Pichon, C.; Gorges, G.; Faure, T.; Boussard, H. *Anaqualand 2.0: Freeware of distances calculations with Frictions on a Corridor*; Cemagref Antony. 2006. Available online: https://www6.rennes.inra.fr/sad/Outils-Produits/Outils-informatiques/Anaqualand (accessed on 10 May 2020).

56. De Mas, F.B. Chapitre III: Barrage mobile à pont supérieur. In *Cours de Navigation Intérieure de l'Ecole Nationale des Ponts et Chaussées*; Rivières Canalisées, Librairie Polytechnique: Paris, France, 1903; pp. 131–143.

57. De Guibert, P. Impact des aménagements pour la navigation sur les niveaux d'eau de la Seine en aval de Paris. *La Houille Blanche* **1997**, *8*, 48–50. [CrossRef]

58. Larinier, M. Upstream and downstream fish passage experience in France. In *Fish Migration and Fish Bypasses*; Jungwirth, M., Schmutz, S., Weiss, S., Eds.; Fishing News Books: Oxford, UK, 1998; pp. 127–145.

59. Katopodis, C.; Williams, J.G. The development of fish passage research in a historical context. *Ecol. Eng.* **2012**, *48*, 8–18. [CrossRef]

60. Lestel, L.; Carré, C. *Les Rivières Urbaines et Leur Pollution*; Collection Indisciplines, Editions Quæ: Versailles, France, 2017; 286p.

61. Romero, E.; Le Gendre, R.; Garnier, J.; Billen, G.; Fisson, C.; Silvestre, M.; Riou, P. Long-term water quality in the lower Seine: Lessons learned over 4 decades of monitoring. *Environ. Sci. Policy* **2016**, *58*, 141–154. [CrossRef]

62. Arrignon, J. Comportement de l'espèce "*Salmo trutta*" dans le bassin de la Seine Suite (1). *Bull. Fr. Piscic.* **1968**, *228*, 77–101. [CrossRef]

63. Lansing, J.S. Complex adaptive systems. *Annu. Rev. Anthropol.* **2003**, *32*, 183–204. [CrossRef]

64. Meybeck, M.; Lestel, L. A Western European River in the Anthropocene: The Seine, 1870–2010. In *Rivers of the Anthropocene*; Kelly, J.M., Scarpino, P., Berry, H., Syvitski, J., Meybeck, M., Eds.; University of California Press: Oakland, CA, USA, 2017; pp. 84–100.

65. Soetaert, K.; Middelburg, J.J.; Heip, C.; Meire, P.; Van Damme, S.; Maris, T. Long-term change in dissolved inorganic nutrients in the heterotrophic Scheldt estuary (Belgium, The Netherlands). *Limnol. Oceanogr.* **2006**, *51*, 409–423. [CrossRef]

66. Fernandes, M.R.; Aguiar, F.; Martins, M.; Rivaes, R.; Ferreira, M. Long-term human-generated alterations of Tagus River: Effects of hydrological and land-use changes in distinct river zones. *Catena* **2020**, *188*, 1–14. [CrossRef]

67. Winiwarter, V.; Schmid, M.; Dressel, G. Looking at half a millennium of co-existence: The Danube in Vienna as a socio-natural site. *Water Hist.* **2013**, *5*, 101–119. [CrossRef]

68. Limburg, K.E.; Waldman, J.R. Dramatic declines in North Atlantic diadromous fishes. *BioScience* **2009**, *59*, 955–965. [CrossRef]

69. Lavollée, G. Le saumon en Seine. *Bull. Soc. Cent. Aquic. Pêche* **1902**, *14*, 221–234.

70. Tales, E.; Jérôme, B.; Beslagic, S.; Stefani, F.; Tartari, G.; Wolter, C. Réponse des Peuplements de Poissons à L'urbanisation et aux Altérations Anthropiques à Long Terme des Cours d'eau. In *Les Rivières Urbaines et Leur Pollution*; Lestel, L., Carré, C., Eds.; Collection Indisciplines, Editions Quæ: Versailles, France, 2017; pp. 242–252.

71. Maes, J.; Stevens, M.; Breine, J. Modelling the migration opportunities of diadromous fish species along a gradient of dissolved oxygen concentration in a European tidal watershed. *Estuar. Coast. Shelf Sci.* **2007**, *75*, 151–162. [CrossRef]

72. Moreau, E. Les poissons du département de l'Yonne. *Bull. Soc. Sci. Hist. Nat. Yonne* **1898**, *52*, 3–82.

73. Garnier, J.; Billen, G.; Cébron, A. Modelling nitrogen transformations in the lower Seine river and estuary (France): Impact of wastewater release on oxygenation and N^2O emission. *Hydrobiologia* **2007**, *588*, 291–302. [CrossRef]

74. Euzenat, G.; Penil, C.; Allardi, J. *Migr'en Seine. Stratégie Pour le Retour du Saumon en Seine*; Rapport Conseil Supérieur de la Pêche/SIAAP: Paris, France, 1992; 38p.

75. Guillard, J.; Colon, M. First results on migrating shad (*Alosa fallax*) and mullet (*Mugil cephalus*) echocounting in a lock on the Rhône River (France) using a split-beam sounder, and relationships with environmental data and fish caught. *Aquat. Living Resour.* **2000**, *13*, 327–330. [CrossRef]

76. Garcier, R. Rivers we can't bring ourselves to clean–historical insights into the pollution of the Moselle River (France), 1850–2000. *Hydrol. Earth Syst. Sci.* **2007**, *11*, 1731–1745. [CrossRef]

77. Van Dijk, G.; Marteijn, E.; Schulte-Wülwer-Leidig, A. Ecological rehabilitation of the River Rhine: Plans, progress and perspectives. *Regul. Rivers Res. Manag.* **1995**, *11*, 377–388. [CrossRef]

78. Roy, M.L.; Le Pichon, C. Modelling functional fish habitat connectivity in rivers: A case study for prioritizing restoration actions targeting brown trout. *Aquat. Conserv. Mar. Freshw. Ecosyst.* **2017**, *27*, 927–937. [CrossRef]

79. Beier, P.; Majka, D.R.; Spencer, W.D. Forks in the road: Choices in procedures for designing wildland linkages. *Conserv. Biol.* **2008**, *22*, 836–851. [CrossRef] [PubMed]

80. Le Pichon, C.; Coustillas, J.; Zahm, A.; Bunel, M.; Gazeau-Nadin, C.; Rochard, E. Summer use of the tidal freshwaters of the River Seine by three estuarine fish: Coupling telemetry and GIS spatial analysis. *Estuar. Coast. Shelf Sci.* **2017**, *196*, 83–96. [CrossRef]

81. Baisez, A.; Bach, J.-M.; Leon, C.; Parouty, T.; Terrade, R.; Hoffmann, M.; Laffaille, P. Migration delays and mortality of adult Atlantic salmon *Salmo salar* en route to spawning grounds on the River Allier, France. *Endanger. Species Res.* **2011**, *15*, 265–270. [CrossRef]

82. Haidvogl, G.; Winiwarter, V.; Brumat, S. Natural History of the Danube Region. In *Danube: Future Interdisciplinary School Proceedings 2017*; Hanus, C., Steiner, G., Eds.; Edition Donau Universitat Krems: Krems an der Donau, Austria, 2019; pp. 43–54.

83. Hein, T.; Funk, A.; Pletterbauer, F.; Graf, W.; Zsuffa, I.; Haidvogl, G.; Schinegger, R.; Weigelhofer, G. Management challenges related to long-term ecological impacts, complex stressor interactions, and different assessment approaches in the Danube River Basin. *River Res. Appl.* **2019**, *35*, 500–509. [CrossRef]

84. Bouleau, G. La catégorisation politique des eaux sous l'angle de la political ecology: Le patrimoine piscicole et la pollution en France. *Lespace Geogr.* **2017**, *46*, 214–230. [CrossRef]

85. Violette, A. La disparition du saumon. *Bull. Soc. Cent. Aquic. Pêche* **1902**, *14*, 181–197.

86. Boet, P.; Duvoux, B.; Allardi, J.; Belliard, J. Incidence des orages estivaux sur le peuplement piscicole de la Seine à l'aval de l'agglomération parisienne (Bief Andresy-Mericourt). *La Houille Blanche* **1994**, 141–147. [CrossRef]

87. Moore, H.E.; Rutherfurd, I.D. Lack of maintenance is a major challenge for stream restoration projects. *River Res. Appl.* **2017**, *33*, 1387–1399. [CrossRef]

88. Wang, S.; Flipo, N.; Romary, T. Assimilating high-frequency data in a hydro-biogeochemical model of river systems, the ProSe-PA approach. *Geophys. Res. Abstr.* **2019**, *21*, 1.

89. Ducharne, A.; Eric, S.; Habets, F.; Deque, M.; Gascoin, S.; Hachour, A.; Martin, E.; Oudin, L.; Page, C.; Terray, L.; et al. Potential evolution of the Seine River flood regime under climate change. *La Houille Blanche* **2011**, *1*, 51–57. [CrossRef]

Article

How Do Eutrophication and Temperature Interact to Shape the Community Structures of Phytoplankton and Fish in Lakes?

Liess Bouraï [1,2,*]**, Maxime Logez** [1,2]**, Christophe Laplace-Treyture** [2,3] **and Christine Argillier** [1,2]

[1] Equipe Fonctionnement et Restauration des Hydrosystèmes Continentaux (FRESHCO), Risques, Ecosystèmes, Vulnérabilité, Environnement, Résilience (UR RECOVER), Institut National de la Recherche Pour L'agriculture, L'alimentation et L'environnement (INRAE), Aix Marseille University, F-13100 Aix-en-Provence, France; maxime.logez@inrae.fr (M.L.); christine.argillier@inrae.fr (C.A.)

[2] Pôle R&D Ecosystèmes Lacustres (ECLA), F-13100 Aix-en-Provence, France; christophe.laplace-treyture@inrae.fr

[3] UR EABX (Ecosystèmes Aquatiques et Changements Globaux), INRAE (Institut National de la Recherche Pour L'agriculture, L'alimentation et L'environnement), F-33612 Cestas, France

* Correspondence: liess.bourai@inrae.fr

Received: 30 January 2020; Accepted: 9 March 2020; Published: 11 March 2020

Abstract: Freshwater ecosystems are among the systems most threatened and impacted by anthropogenic activities, but there is still a lack of knowledge on how this multi-pressure environment impacts aquatic communities in situ. In Europe, nutrient enrichment and temperature increase due to global change were identified as the two main pressures on lakes. Therefore, we investigated how the interaction of these two pressures impacts the community structure of the two extreme components of lake food webs: phytoplankton and fish. We modelled the relationship between community components (abundance, composition, size) and environmental conditions, including these two pressures. Different patterns of response were highlighted. Four metrics responded to only one pressure and one metric to the additive effect of the two pressures. Two fish metrics (average body-size and biomass ratio between perch and roach) were impacted by the interaction of temperature and eutrophication, revealing that the effect of one pressure was dependent on the magnitude of the second pressure. From a management point of view, it appears necessary to consider the type and strength of the interactions between pressures when assessing the sensitivity of communities, otherwise their vulnerability (especially to global change) could be poorly estimated.

Keywords: global change; nutrients; anthropogenic pressure stressor; interaction; multiple stressors; lake systems

---

1. Introduction

Freshwater ecosystems are among the systems most threatened and impacted by anthropogenic activities [1]. They are characterized by their high biodiversity [2], the erosion of which is considered to be steeper than that of terrestrial ecosystems [3], which makes them more vulnerable. Aquatic ecosystems are exposed to numerous anthropogenic stressors, be they physical (i.e., habitat degradation), chemical, or biological (i.e. invasive species), which interact with global change and lead to additional perturbations [4–8]. These multiple stressors compromise freshwater biodiversity and its associated biological functions, and ultimately the services provided by these systems to our society [9,10].

In Europe, monitoring associated with the implementation of the Water Framework Directive (WFD; an environmental policy that aims to protect and restore continental aquatic systems) showed that numerous aquatic ecosystems are impaired by human activities, with their ecological status ranging from bad to moderate [4,11]. While the WFD calls for lakes (like all bodies of water) to be in 'good ecological status', the latest evaluation (2nd River Basin Assessment) showed that 45% of lakes did not reach this status [12]. Unfortunately, in certain hydrographic basins, this percentage may even reach 100%. The report of the European Environmental Agency [12] emphasized that lakes are particularly impacted by nutrient enrichments, and also by climate change, more specifically by temperature increases [4,12,13]. The impact of these two stressors is also the most studied. In their review based on 219 studies, Nõges et al. [10] revealed that 78% dealt with nutrient impacts and 31% with temperature effects, either solely or jointly.

Nevertheless, although lakes rarely face a single pressure, few studies have looked at the combined effects of several pressures, and even fewer have examined their interactions. Some results concerned the cumulative effects of temperature and eutrophication on phytoplankton. Both pressures affect phytoplankton abundance and composition [14–17]. For instance, Krosten et al. [18] showed that temperature and nutrients both increase the relative abundance of cyanobacteria. However, in some cases, related by Richardson et al. [19], this relation can be different, depending on the lake type (morphological characteristics of the lake). Similarly, Jeppesen et al. [20] observed that fish community composition changed as a result of these two pressures. Coldwater species were replaced by warm-water-tolerant species owing to longer warm periods in summer and lower oxygen concentrations.

In most of these studies, when multi-stressors were considered, their combined effect was commonly assumed to be additive, i.e., equal to the sum of the individual effects of the stressors acting in isolation. This additive model is increasingly discussed in ecological systems in terms of antagonistic and synergistic interactions [4,7,10,21,22]. Stressors can act in synergy when the combined effect of stressors is greater than the sum of the impacts of individual stressors, whereas antagonistic interactions occur when the combined effect of stressors is less than expected based on their individual effects [23]. In situ, temperature and nutrient enrichment interaction is very often assumed to impact biological communities, but it is rarely quantified or modelled (see Rigosi et al. [24] for exceptions). This phenomenon is better studied in experimental conditions under which nutrient concentrations and temperature can be controlled [21,25–27]. Nevertheless, from a management point of view, the value of a study of this kind could be limited (e.g., the discrepancy between scales, other factors involved, etc.), and without taking into account these interactions (if verified), the evaluation of the lake status could be biased (e.g., [28]). In addition, knowledge of these interactive effects can be useful in the implementation of management plans, with the ecological benefit resulting from efforts to reduce interactive multi-stressors possibly giving rise to some 'ecological surprises' [5,29,30].

The aim of this study was to assess whether: (i) temperature and eutrophication impact various components (metrics) of the lake community, such as productivity and size structure, (ii) whether these effects are additive, (iii) or whether they are multiplicative, i.e., whether the effect of one pressure depends on the other pressure. We focused on two biological groups at the two ends of the trophic chain that are representative of the lake community, are often used in bioindication, and are studied in multi-stress conditions: phytoplankton and fish [7,10,31–34]. Moreover, a given pressure could impact each biological element either directly or indirectly through cascade effects along the trophic chain [17,35–37]. Hence, we could hypothesize similar responses to pressure from both trophic levels expressed by the increase in primary productivity and, thus, fish productivity with temperature and nutrients [37–40]. Similarly, we expected change to the community and a negative relationship between temperature and sizes [35,41–43].

This study was conducted at the macro-ecological scale, on a consistent dataset of 204 French lakes to ensure large diversity in thermal and trophic conditions. The combined effect of temperature and eutrophication was studied by comparing three statistical models: one considering only the effect of lake morphology, a second model considering an additive effect of the two pressures, and a more complex model considering the interaction between pressures.

2. Materials and Methods

2.1. Biological Data

The dataset comprised 204 French lakes, 48 natural lakes, and 156 reservoirs, for which biological and environmental data were available and collected in a standardized manner.

Fish samples were collected according to the Norden gillnet standardized protocol [44] during the period between July and mid-October. This protocol is based on a randomly stratified sampling design. Benthic gillnets, 30 m in length, 1.5 m in height, composed of 12 panels with mesh sizes ranging from 5 to 55 mm knot-to-knot, were randomly distributed in the depth strata of the lakes. The gillnets were set before sunset and lifted after sunrise to cover peaks of maximal fish activity [45,46]. All fish caught were identified at the species level, then measured (total length in millimeters) and weighed (to the nearest gram).

Phytoplankton was collected using a standardized method [47] and processed in laboratory following the counting process of the European Standard NF15204 [48]. Four sampling campaigns a year are recommended for each lake: three during the warmer period (between May and October) and one in late winter. Phytoplankton was collected at the deepest point of the lakes, in the euphotic part of the water column. Taxa were determined at the species level in the laboratory and their abundances were weighted by taxa biovolume [49] using standard cell values defined in the software Phytobs [50], or measured directly from the sample if the values were lacking. Additionally, chlorophyll-a was collected from the euphotic zone during each sampling event and measured using the standard methods NF-T 90-117 [49,51,52].

2.2. Biological Characterization

For each sampling occasion, ichthyofauna was characterized by eight metrics related to its density, by its composition, and by the size of the individuals making up the communities. Density was estimated by the number of individuals caught per sampling unit effort (expressed in square meters of nets set during a 12-h night period; CPUE) and the total biomass caught per sampling unit effort (BPUE). The ratio between the abundance of a predator, the perch (*Perca fluviatilis*), and the abundance of a prey, the roach (*Rutilus rutilus*), was used as a proxy of the trophic equilibrium of the lakes [53,54]. These two species are very common in French temperate lakes and generally abundant [55,56]. BPUEs and CPUEs were calculated for both species and their ratios (BPUE_Perch/Roach and CPUE_Perch/Roach) were computed for all the lakes where the two species occurred. In addition, the ratio of average perch to roach body size (Average Perch/Roach Body Size) was calculated for each lake to measure the evolution of ichthyofauna composition. The overall size of the fish community was assessed by the average size of all the fish caught in the benthic gillnets in a lake. This metric is useful for comparing the average difference in fish size between communities without differentiating between the processes behind it (loss of the largest individuals, decrease in the size of all fish, increase of small species or individuals, etc.) [57,58]. To investigate the processes involved in the change in size structure, the community size spectra (CSS) were considered [59]. CSS represent a frequency distribution of individual body sizes across size classes (defined on a log-scale) irrespective of taxonomy, through a linear regression relating abundances to size classes on log scales. Two metrics were calculated: the midpoint and the slope of the regression. The midpoint (CSS_Midpoint) value is an indicator of productivity of the system and determines the level of richness of ecosystems. For instance, two communities can display the same slope but different midpoints if the one has more fish than the other [60]. Slope (CSS_Slope) is an indicator of the health of the community [60], for example, fish overexploitation will reduce the abundance of large fish traduced by a high slope value.

Four metrics relating to density, composition, or size of phytoplankton communities were considered. Phytoplankton total biomass was surrogated by the concentration of chlorophyll-a (Chl-a, µg/L) in the euphotic zone. To limit the impact of seasonal variability, the Chl-a concentrations of the three summer samples were averaged. The composition of phytoplankton was assessed by the abundances of cyanobacteria and golden alga (Chrysophytes). Cyanobacteria are assumed to benefit from warmer temperatures and become dominant in higher nutrient concentrations, whereas Chrysophyceae are assumed to prefer colder and less eutrophic conditions [41,61–63]. Their abundances were weighted by taxa biovolume (considered as fixed) in order to calculated their respective biomass (Cyano_Biovolume and Chryso_Biovolume), and expressed in cubic millimeters per liter (mm^3/L). We then defined two size classes of phytoplankton, each taxon was classified as large or small, irrespective of whether the taxon was considered larger or smaller than microphytoplankton, as defined in the literature [64]. The ratio of the biomasses of large taxa on the biomasses of small taxa in each lake was then calculated. This metric (phytoplankton size class) allows us to see, as with ichthyofauna, whether the size structure of the phytoplankton community is affected by thermal and/or eutrophic stress [41,42,62,65,66].

2.3. Environmental Data

The lakes were characterized by natural environmental variables potentially influencing the structure of biological assemblages [67,68], i.e., physical/morphological characteristic of their environment, and by stressors.

The dataset comprised 204 French lakes, 48 natural lakes and 156 reservoirs. This corresponds to a diversity of lake throughout the French territory with morphological characteristics, ranging from the plain lake to the mountain lake, from the shallow lake to the deep lake, and very varied in terms of surface, area, the shape of the lake basin, and mean temperature or trophy level (details of the calculation of the index below) (Table 1, see Supplementary Materials for more details).

Table 1. Characterization of the environmental variables of the lakes studied.

Environmental Variable	Minimum–Maximum	Mean (SD)
Altitude (m)	0–2061	404.8 (382.4)
Perimeter (m)	1.5×10^3–1.9×10^5	1.7×10^4 (2.1×10^4)
Depth (m)	0.3–154.2	12.7 (16.6)
Area (km^2)	0.1–577.1	6.1 (40.9)
Volume (m^3)	1.3×10^5–8.9×10^{10}	5×10^8 (6.2×10^9)
Ig (m/km)	0.1–155.7	8.1 (16.8)
Temperature (°C)	2.9–16.3	11.1 (2.2)
Eutrophication	−2.0–3.1	−0.01 (1.1)

In order to limit the multi-collinearity between predictors (thus the redundancy between physical variables), we ran a principal component analysis. Four variables emerged from this analysis and characterized lake morphology: the mean depth (Depth, m), the area (Area, km^2), the water volume (Volume, m^3), and the overall hill index (Ig, m/km). Ig was calculated from the maximum depth, perimeter, and area [69,70] that characterizes the shape of the lake basin, according to the following Equation (1):

$$I_g = \frac{Pmax}{Lr},\tag{1}$$

with

$$Lr = Plake \times \frac{0.282}{1.128} \times \left(1 + \sqrt{1 - \left(\frac{1.128}{K_C}\right)^2}\right),\tag{2}$$

and

$$K_C = 0.282 \times \frac{Plake}{\sqrt{Alake}},\tag{3}$$

where *Pmax* is the maximum depth of the lake (m), *Plake* is its perimeter (km) and *Alake* its area (km^2).

Two stressors were considered: temperature and eutrophication. Because water temperatures were not available for all the lakes but are strongly correlated with air temperatures [71,72] ($R^2 = 0.82$), we used the latter to characterize lake temperatures. To this end we used data from the SAFRAN reanalysis [73,74], available at a spatial resolution of 8 km × 8 km. To integrate the difference of altitude between grid cells and lakes, a correction of 6.5×10^{-3} C/m was applied.

Eutrophication was described by three variables: total phosphorous concentration (TP, µg/L), nitrate concentration (NO$_3$, µg/L) and importance of non-natural land cover in the catchment area (NNLC, percentage of the total catchment area). Phosphorus values range from 5 µg/L to 464 µg/L, corresponding to trophic states classified from oligotrophic to hyper-eutrophic [75]. The nutrients were sampled in the euphotic zone during the four same annual campaigns as for phytoplankton, according to a standard sampling method. NNLC was defined as the percentage of non-natural areas in the catchment and derived from the Corine Land Cover database [76]. This encompassed the CLC categories: (1) artificial territories and (2) agricultural territories (without 23 grasslands) [77]. We summarized the information of these eutrophication measures in a synthetic index of eutrophication. First, TP and NO$_3$ were log-transformed, NNLC was transformed by the arcsin of the square root, and then each variable was centered and reduced. The three transformed variables were averaged and these values were centered and reduced to produce the synthetic index.

All the data used in this study were collected between 2005 and 2017 by research institutes or water agencies, and centralized in a database by our laboratory.

To reduce the skewness of their distribution we log transformed maximum depth, lake area, lake volume, Ig, BPUEs, CPUEs, Chl-a, and the biovolumes of cyanobacteria and Chrysophyceae.

2.4. Data Analysis

To assess the significance of the interactions of pressures on biological metrics we defined three nested linear models related to three hypotheses [78]: (i) no pressure effect, (ii) additive effect of pressures, and (iii) interaction of pressures (multiplicative effect of pressures). The first model related the variability of biological metrics to physical variables only, metric ~ depth + area + Volume + Ig corresponds to the environmental data block (called 'environment' in the next formulae); the second model integrated the physical variables and the variable of pressures in an additive manner, metric ~ environment + temperature + eutrophication; the more complex model integrated the interaction between temperature and eutrophication, metric ~ environment + temperature + eutrophication + temperature interacting with eutrophication. The first model assumed that biological metric variability depends only of the environmental conditions. The second model hypothesizes that the effect of each pressure is independent of the other. In other words, whatever the level of the second pressure, the magnitude of response to the first pressure will always be the same. Conversely, the interaction included in the third model assumed that the effect magnitude of one pressure depends on the intensity of the second pressure.

The effect of pressures and their behavior (additive or multiplicative) were tested on each fish and phytoplankton metric by comparing models two by two with ANOVA (*F*-tests). First, we tested model 3 vs. model 2, then, if the interaction was not significant, we tested model 2 vs. model 1 to verify the significance of the pressure effect on the metric variability (*F*-tests). Once the most explanatory model was selected, we visually checked whether the linear model assumptions were verified (i.e., homoscedasticity, normality of residuals). Only metrics for which more than 10% of the variability of the biological metrics was explained were retained.

Finally, because it would have been difficult to forecast from coefficient values, if the interaction was significant, we looked at its effect in graph form using graph effect display representation [79]. For each graph, we represented how the expected metric values varied along the pressure gradients by leaving pressure values and fixing the values of the environmental variable to their averages. For each metric, two sets of graphs were drawn. One graph was compiled by allowing pressure values vary across their observed range of values, and one graph was drawn by restricting pressures to their observed combination of values.

3. Results

3.1. Environment and Pressures

The two metrics describing the pressures were weakly correlated with each other and with the natural environmental variables (Table 2). Temperature varied between 2.9 °C and 16.3 °C and eutrophication between −2 (low level of eutrophication) and 3.1 (high level of eutrophication) (Table 1). Not all possible combinations of pressure values were observed (white space in the right lower part of Figure 1). For instance, no lakes with an index value of eutrophication greater than −0.5 had a temperature lower than 8 °C, or an eutrophication greater than 1 with a temperature lower than 8 °C. Finally, only a few were present at a temperature below 6 °C. These limitations of our dataset conditions will be taken into account in the following interaction analyze of pressure effects on biological metrics.

Table 2. Pearson correlation of lake characteristics with temperature and eutrophication pressures.

Environmental Variable	Temperature	Eutrophication
Depth (m)	−0.22	−0.43
Area (km^2)	0.14	−0.21
Volume (m^3)	−0.01	−0.37
Ig	−0.39	−0.41
Temperature(°C)	−	0.32
Eutrophication	0.32	−

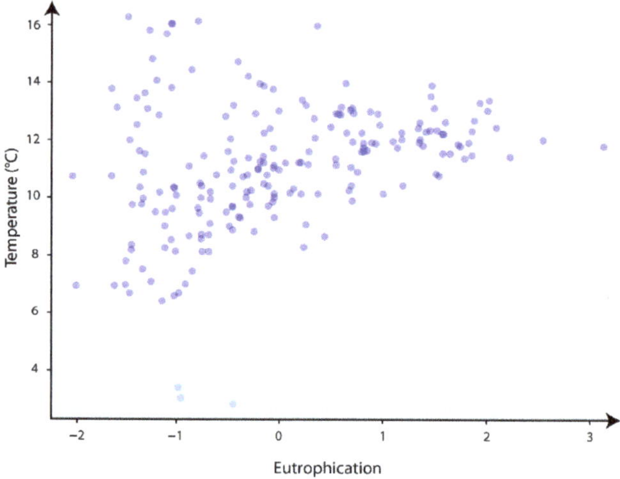

Figure 1. Relationship between temperature and eutrophication of lakes.

3.2. Pressure Effects

Of the 12 metrics, four (CSS_Slope, Phytoplankton Size Class, Average Perch/Roach Body Size, Cyano_Biovolume) were not sufficiently explained by the environmental and pressure variables ($R^2 < 10\%$) and were not considered further.

One model (CPUE_Perch/Roach) with the only environmental effect, five models (BPUE, CPUE, CSS_Midpoint, Chl-a, Chryso_Biovolume) with an additive effect of pressure and two models (Average Fish Body Size, BPUE_Perch/Roach) with a significant interaction of pressures were selected (Table 3).

Table 3. Model comparisons two by two (ANOVA) with the final model selected and its associated adjusted R^2 ('-' if the R^2 was lower than 0.1). F-test results: *** $p < 0.001$, ** $p < 0.01$ and * $p < 0.05$. BPUE, biomass caught per sampling unit effort; CPUE, catch per sampling unit effort; CSS, community size spectra; Chl-a, chlorophyll-a; Cyano, cyanobacteria; Chryso, Chrysophyceae.

Metric	Model 2 vs. Model 3	Model 1 vs. Model 2	Selected Model	R^2
BPUE	$F_{1,196} = 1.06$	$F_{1,197} = 11.86$ ***	Model 2	0.47
CPUE	$F_{1,196} = 0.88$	$F_{1,197} = 10.67$ ***	Model 2	0.32
BPUE_Perch/Roach	$F_{1,174} = 6.69$ **	$F_{1,175} = 1.11$	Model 3	0.22
CPUE_Perch/Roach	$F_{1,174} = 2.58$	$F_{1,175} = 0.70$	Model 1	0.15
Average Perch/Roach Body-Size	$F_{1,174} = 0.27$	$F_{1,175} = 0.91$	–	0
Average Fish Body Size	$F_{1,196} = 8.30$ **	$F_{1,197} = 5.72$ **	Model 3	0.12
CSS_Midpoint	$F_{1,196} = 0$	$F_{1,197} = 4.40$ *	Model 2	0.26
CSS_Slope	$F_{1,196} = 0.35$	$F_{1,197} = 0.02$	–	0.04
Chl-a	$F_{1,196} = 0.19$	$F_{1,197} = 20.32$ ***	Model 2	0.41
Cyano_Biovolume	$F_{1,196} = 2.06$	$F_{1,197} = 1.60$	–	0.09
Chryso_Biovolume	$F_{1,196} = 0.28$	$F_{1,197} = 9.90$ ***	Model 2	0.10
Phytoplankton size class	$F_{1,196} = 0.27$	$F_{1,197} = 4.57$ *	–	0.07

When an additive effect of pressure was significant, eutrophication was always positively related to the metric values (positive coefficient; Table 4), with the exception of Chryso_Biovolume, which decreased with eutrophication (Table 4). Fish metrics influenced only by eutrophication pressure were BPUE and CSS_Midpoint. The R^2 value indicated that the models explained 47% and 26% of variability, respectively. When the eutrophication index was removed from these models, the explained variance decreased by 6% and 3%. Metrics of phytoplankton influenced only by eutrophication were Chl-a and Chryso_Biovolume, with explained variances of 41% and 10%, respectively. Compared with the environmental model (model 1), including the eutrophication index, the explained variability increased by 11% and 8%, respectively. Fish CPUE was the only metric significantly influenced by the additive effect of the two pressures considered and with positive relationships. The model explained 32% of the variability of the metric (Table 3) and 7% of the variance was explained only by the combined effect of pressures.

Table 4. Model coefficient (positive + or negative −) of pressure (temperature, eutrophication and interaction) for biological metrics selected. Bold: the impact of pressure or interaction on metric is significant.

Metric	Temperature	Eutrophication	Interaction
BPUE	+0.03	**+0.14**	No interaction
CPUE	**+0.06**	**+0.22**	No interaction
BPUE_Perch/Roach	**−0.17**	**+1.91**	**−0.16**
Average Fish Body-Size	**−0.43**	**−50.56**	**+4.04**
CSS_Midpoint	+0.04	**+0.13**	No interaction
Chl-a	+0.05	**+0.43**	No interaction
Chryso_Biovolume	+0.06	**−0.90**	No interaction

3.3. Interaction of Pressures

A significant negative effect of the interaction between eutrophication and temperature was measured on the BPUE_Perch/Roach metric, but a positive effect of the interaction between these two pressures was measured on the Average Fish Body Size metric. In these models, BPUE_Perch/Roach was negatively related to temperature and positively to eutrophication. Average Fish Body Size was negatively related to both temperature and eutrophication (Table 4). The interactive models explained 22% and 12% of the variability of the BPUE_Perch/Roach and Average Fish Body Size metrics, respectively. Compared with the R^2 value of the additive model (19% and 9%), the gain in explained variability relative to the interaction model represented an increase of 3% for both.

The interaction effects between temperature and eutrophication on BPUE_Perch/Roach and Average Fish Body Size metrics were assessed graphically (Figures 2 and 3). In the case of

BPUE_Perch/Roach, we observed an interval of values approximately three times lower at low levels of eutrophication than at high levels, which means a higher effect of temperature at high eutrophication levels (Figure 2b). This corresponds to a small increase in BPUE_Perch/Roach with temperature for low levels of eutrophication and a large decrease with temperature at high levels of eutrophication. At low temperatures (<10–12 °C) we observed higher BPUE_Perch/Roach values than at higher temperatures (Figure 2c), which is accompanied by an increase in the metric with eutrophication at low temperatures and a decrease at high temperatures.

By looking at the interaction only on the combination of pressure values observed in lakes, the magnitude of response was limited (Figure 2d). We saw a small increase in the ratio of temperature to the low level of eutrophication (Figure 2e) and a significant decrease with the temperature for high levels of eutrophication (Figure 2f), reaching lower ratio values than at low temperatures. The high values of the BPUE_Perch/Roach ratio visible on the full model (Figure 2a) at low temperature–high eutrophication were not visible with the *in situ* pressure values (Figure 2d).

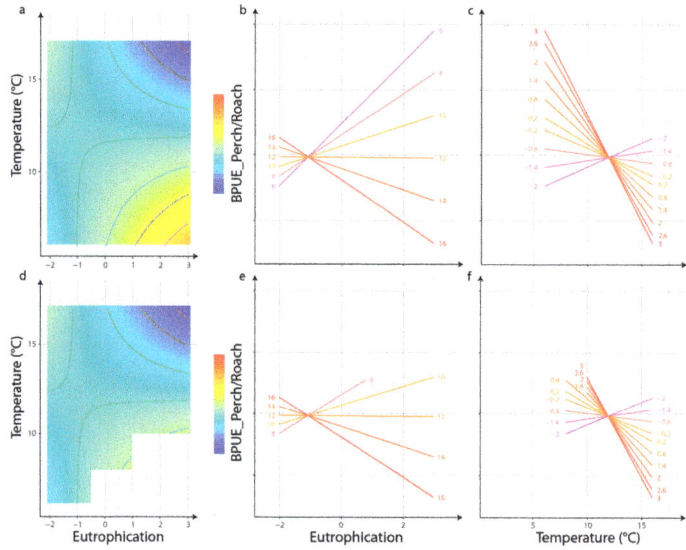

Figure 2. Effect of interaction between average temperature and eutrophication level on BPUE_Perch/Roach metric (log) when considering all possible combinations of pressures (**a–c**), or when considering only the observed combination of pressures (see Figure 1) (**d–f**). (**a,d**) Low theoretical values are represented in blue and high theoretical values in red.

When temperature increased, Average Fish Body Size decreased at low eutrophication levels, and increased strongly at high levels of eutrophication (Figure 3b). We observed higher Average Fish Body Size at low eutrophication, whereas we observed the lowest values in high eutrophication and low temperature conditions. At low temperatures (2–12 °C), body size varied widely with eutrophication and the highest values were observed when eutrophication was low (Figure 3c). Conversely, when temperatures were high (12–17 °C), the highest values of fish body size were measured when eutrophication was significant. Average Fish Body Size decreased with eutrophication at low temperatures and increased with eutrophication at high temperatures.

When analysis was limited to the pressure values observed, we essentially detected an increase in size with eutrophication for high temperatures (>13 °C) (Figure 3e) and a decrease in body size with temperature for low levels of eutrophication (Figure 3f). Compared with Figure 3a, the amplitude of body size in response to pressure conditions was reduced. The lowest values associated with an increase in temperature at low eutrophication were not observed with the observed pressure values (Figure 3d).

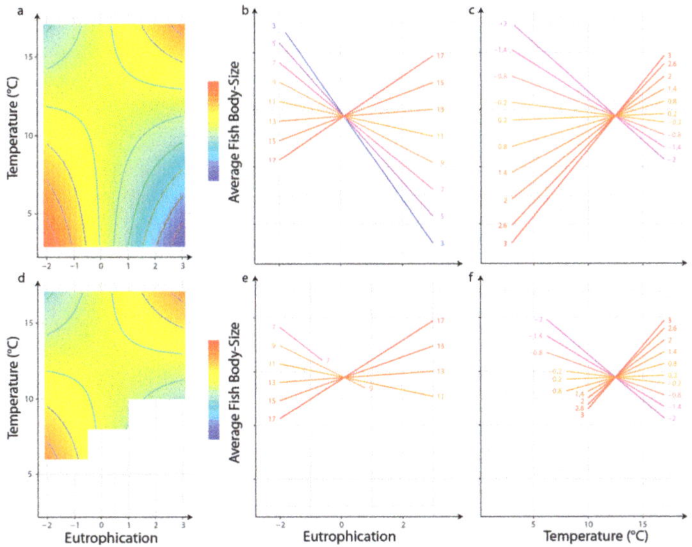

Figure 3. Effect of interaction between average temperature and eutrophication level on Average Fish Body Size when considering all possible combinations of pressures (**a–c**), or when considering only the observed combination of pressures (see Figure 1) (**d–f**). (**a,d**) Low theoretical values are in blue and high theoretical values in red.

4. Discussion

The objective of our study was to assess the interactive effect of temperature and eutrophication on the structure of fish and phytoplankton communities. Among the twelve pressure/impact models developed, an additive effect and an interactive effect were detected for, respectively one and two fish metrics, while most of the models reveal a significant effect of one stressor.

The impact of eutrophication on biological communities has long been observed [41]. For example, the effects of phosphorus loadings on primary production have largely been described in the scientific literature [80–82]. Algal blooms in response to eutrophication are also well-documented [83], as well as the changes in community structure [34,42,66,84,85]. The impact of temperature has been explored in detail, and is often still studied, especially since climate change has become evident [86]. The effect of an increase in temperature could be manifold and complex (see, for instance, Keller [87] and Richardson et al. [19]), but many authors agree on an increase in productivity [38,88,89] or on a decrease in ectothermal size [57,90,91]. Most of our results are in accordance with these observations. Fish density expressed in occurrence (CPUE) was shown to be positively correlated with an increase in both temperature and eutrophication. Similarly, the biomass of fish per capture effort (BPUE) and Chl-a were positively related to nutrient enrichment. This increasing productivity of phytoplankton and fish with eutrophication [36,38,92] is generally associated with the shift in community composition and structure [35], which is also observed in our case, through the ratio of perch vs. roach biomasses and Chrysophytes biomass. The biomass of Chrysophytes was shown to decrease when eutrophication increases, which is consistent with our hypothesis and previous results [34,35,85].

Of the two metrics related to CSS, slope and midpoints, only the latter was shown to increase with eutrophication. Finally, four metrics for which response to eutrophication and/or temperature were expected, were not explained by our models: CSS_Slope, Phytoplankton Size Class, Average Perch/Roach Body Size, and Cyano_Biovolume. The absence of impact of the stressors can be attributed to sampling protocol (reduction of the size range variability by gillnet selectivity for fish) and to size assessment for phytoplankton (very simplified and coarse) [41,42,66]. In addition, the high temporal and spatial variability of the abundance of cyanobacteria is a limit to this type of analysis [93,94].

In addition to the effect of pressures, we saw the significant proportion of model variability explained by the environmental characteristics of lakes confirming previous results and patterns when focusing on pressures [67,68].

More interestingly, two metrics were sensitive to the interaction of temperature and eutrophication: BPUE ratio between perch and roach and average community size. The interaction of these two pressures on the BPUE ratio between perch and roach highlighted the role played by temperature on the magnitude of this relationship. This was even more evident when graph effect displays were represented only on the observed range of values of these pressures (Figure 2d–f). The slope of the relation between eutrophication and BPUE ratio increased as the temperature increased (especially between 12 and 16 degrees). When all possible combinations of the pressures (Figure 2a–c) are used to visualize the estimated effect of each pressure (taking into account the second pressure due to the interaction between the two), some unexpected relationship may appear: for instance, a positive relationship between the ratio of perch/roach biomasses and nutrient enrichment for cold lakes (see Figure 2b). This is probably due to deep extrapolations for non-observed pressure conditions. Unlike an experimental design, for which environmental conditions are controlled and a perfect crossover of pressures can be used, the cold lakes in our dataset were mainly oligotrophic. Eutrophic lakes were predominantly observed under cool and warm conditions (Figure 1). However, interactive effects of temperature and nutrients on community dynamics are very often studied and observed on phytoplankton [37,84,85], but poorly tested for fish.

The relationship between fish size and temperature has been well studied, especially in the context of global warming (e.g., [57]). The significant interaction between eutrophication and temperature suggested that the magnitude as well as the sign of the relationship between temperature and average community size depends on trophic level. In oligotrophic conditions, community size was estimated to decrease along thermal gradients. This pattern has already been observed for fish [57,91], especially in lakes [95]. Ectothermal individuals could be smaller in warmer conditions, according to the temperature size rule theory [90], and/or smaller species could be preferentially selected as temperature increases [57,91]. With nutrient enrichment, the model predicted that average community size would increase with temperature, which is contradictory to the theory prediction. Nonetheless, fish in fisheries would grow faster and larger when the temperature increases, but when they are fed ad libitum [96]. This could also be explained by a more efficient trophic transfer and more available resources amplified through the trophic level with warming, as predicted by metabolic theory in nutrient-replete systems [97].

The fact that some components of community structure are impacted by different pressures and, in particular, their interaction should provide water managers with strong insight. Until recently in Europe, water managers mainly focused on pressure—impact relationships through multi-metric indices [98,99] to assess the ecological status of lakes [100,101] owing to the WFD. Such interaction could influence the scoring values of metrics, then metric index values and ecological assessment, but Miguet et al. [28] evaluated it at a small deviation. More recently, rather than focusing on ecological status, which is a current evaluation, some authors have worked on the vulnerability of lake ecosystems (e.g., [102]). This concept was designed around three components—sensitivity (the degree to which communities are affected, either adversely or beneficially, by pressure), exposure (contact between communities and stressors) and capacity to adapt (the ability of communities to adjust to potential hazards, to take advantage of opportunities or to respond to consequences) [103]—and seems very interesting for anticipating/forecasting lakes that will suffer from global warming. Addressing the vulnerability of communities to multiple stressors appears necessary in order to prevent future alterations in aquatic ecosystems by prioritizing the protection of the most vulnerable structures [104,105]. Our study shows that the sensitivity of communities is modulated both by the level of exposure to pressures and by the coupling of these pressures. If the interaction of pressures is seen as an additive effect, while multiple interactions could occur [7], then the sensitivity of the communities might be inconsistently evaluated, since the actual effect of a pressure would be related to the level of the other pressures. Thus, by ignoring interaction, there is a risk of an unexpected ecological effect by underestimating the effect of pressure, or even concluding that an effect in the opposite direction depends on the exposure level to another pressure [5,6]. This could lead to the adoption of an inappropriate strategy to manage lakes or to not prioritizing management actions for lakes that could actually be much more vulnerable than expected. With the increase of stress on freshwater ecosystems such as lakes, it will be necessary to pursue our monitoring on these systems to study their combined effects with global change and how this will impact aquatic communities [7,106].

To conclude, we highlight in situ interactive effects of eutrophication and temperature on lake fish communities. Therefore, in light of these unexpected effects, future management plans should consider the type and strength of interactions in order to avoid underestimating the vulnerability of these environments [105,107]. Finally, a consideration of pressure interaction in the study of environmental vulnerability could help to identify priorities for action to conserve and restore aquatic environments.

Supplementary Materials: The following are available online at http://www.mdpi.com/2073-4441/12/3/779/s1, Table S1: Linear Model Coefficients of biological metrics, Table S2: Environmental characteristic of the 204 lakes in the dataset.

Author Contributions: Conceptualization, investigation, draft preparation, by L.B., M.L, C.L.-T., and C.A.; methodology, formal analysis by L.B. and M.L. All authors have read and agreed to the published version of the manuscript.

Funding: This research was funded by the Research & Development center "ECLA" and by the South Region (Provence-Alpes-Côte d'Azur) grant number n 2018-05953.

Acknowledgments: The authors are grateful to all those who participated in data collection and management, especially Nathalie Reynaud, Thierry Point, and Thierry Tormos. The authors also thank Pierre Alain Danis for his valuable help with temperature data, Paul Miguet for his advice and for English correction, Isabella Athanassiou and Eric Hernquist.

Conflicts of Interest: The authors declare no conflict of interest.

References

1. Dudgeon, D.; Arthington, A.H.; Gessner, M.O.; Kawabata, Z.-I.; Knowler, D.J.; Lévêque, C.; Naiman, R.J.; Prieur-Richard, A.-H.; Soto, D.; Stiassny, M.L.J.; et al. Freshwater biodiversity: Importance, threats, status and conservation challenges. *Biol. Rev.* **2006**, *81*, 163. [CrossRef]
2. Lundberg, J.G.; Kottelat, M.; Smith, G.R.; Stiassny, M.L.J.; Gill, A.C. So Many Fishes, So Little Time: An Overview of Recent Ichthyological Discovery in Continental Waters. *Ann. Mo. Bot. Gard.* **2000**, *87*, 26. [CrossRef]
3. Sala, O.E. Global Biodiversity Scenarios for the Year 2100. *Science* **2000**, *287*, 1770–1774. [CrossRef]
4. Hering, D.; Carvalho, L.; Argillier, C.; Beklioglu, M.; Borja, A.; Cardoso, A.C.; Duel, H.; Ferreira, T.; Globevnik, L.; Hanganu, J.; et al. Managing aquatic ecosystems and water resources under multiple stress —An introduction to the MARS project. *Sci. Total Environ.* **2015**, *503–504*, 10–21. [CrossRef]
5. Jackson, M.C.; Woodford, D.J.; Weyl, O.L.F. Linking key environmental stressors with the delivery of provisioning ecosystem services in the freshwaters of southern Africa: Environmental stress and ecosystem services. *Geo Geogr. Environ.* **2016**, *3*, e00026. [CrossRef]
6. Ormerod, S.J.; Dobson, M.; Hildrew, A.G.; Townsend, C.R. Multiple stressors in freshwater ecosystems. *Freshw. Biol.* **2010**, *55*, 1–4. [CrossRef]
7. Schinegger, R.; Palt, M.; Segurado, P.; Schmutz, S. Untangling the effects of multiple human stressors and their impacts on fish assemblages in European running waters. *Sci. Total Environ.* **2016**, *573*, 1079–1088. [CrossRef] [PubMed]
8. Wagner, T.; Erickson, L.E. Sustainable Management of Eutrophic Lakes and Reservoirs. *J. Environ. Prot.* **2017**, *8*, 436–463. [CrossRef]

9.	Côté, I.M.; Darling, E.S.; Brown, C.J. Interactions among ecosystem stressors and their importance in conservation. *Proc. R. Soc. B Biol. Sci.* **2016**, *283*, 20152592. [CrossRef] [PubMed]

10.	Nõges, P.; Argillier, C.; Borja, Á.; Garmendia, J.M.; Hanganu, J.; Kodeš, V.; Pletterbauer, F.; Sagouis, A.; Birk, S. Quantified biotic and abiotic responses to multiple stress in freshwater, marine and ground waters. *Sci. Total Environ.* **2016**, *540*, 43–52. [CrossRef]

11.	Poikane, S.; Ritterbusch, D.; Argillier, C.; Białokoz, W.; Blabolil, P.; Breine, J.; Jaarsma, N.G.; Krause, T.; Kubečka, J.; Lauridsen, T.L.; et al. Response of fish communities to multiple pressures: Development of a total anthropogenic pressure intensity index. *Sci. Total Environ.* **2017**, *586*, 502–511. [CrossRef] [PubMed]

12.	Kristensen, P.; Whalley, C.; Zal, F.; Christiansen, T.; Schmedtje, U.; Solheim, A.; Austnes, K.; Kampa, E.; Rouillard, J.; Prchalova, H.; et al. *2018 EEA European Waters Assessment*; European Waters Assessment; Office for Official Publications of the European Union: Copenhagen, Denmark, 2018; Volume 7.

13.	Kristensen, P.; Vanneuville, W. *Climate Change, Impacts and Vulnerability in Europe 2012: An Indicator-Based Report Section 3.3: Freshwater Quantity and Quality*; European Waters Assessment; Office for Official Publications of the European Union: Luxembourg, 2012; Volume 12, pp. 119–127.

14.	Chen, X.; Yang, X.; Dong, X.; Liu, E. Environmental changes in Chaohu Lake (southeast, China) since the mid 20th century: The interactive impacts of nutrients, hydrology and climate. *Limnologica* **2013**, *43*, 10–17. [CrossRef]

15.	Salmaso, N. Interactions between nutrient availability and climatic fluctuations as determinants of the long-term phytoplankton community changes in Lake Garda, Northern Italy. *Hydrobiologia* **2011**, *660*, 59–68. [CrossRef]

16.	Scheffer, M.; van Nes, E.H. Shallow lakes theory revisited: Various alternative regimes driven by climate, nutrients, depth and lake size. *Hydrobiologia* **2007**, *584*, 455–466. [CrossRef]

17.	Shurin, J.B.; Winder, M.; Adrian, R.; Keller, W.; Matthews, B.; Paterson, A.M.; Paterson, M.J.; Pinel-Alloul, B.; Rusak, J.A.; Yan, N.D. Environmental stability and lake zooplankton diversity—Contrasting effects of chemical and thermal variability. *Ecol. Lett.* **2010**, *13*, 453–463. [CrossRef]

18.	Kosten, S.; Huszar, V.L.M.; Bécares, E.; Costa, L.S.; Donk, E.; Hansson, L.-A.; Jeppesen, E.; Kruk, C.; Lacerot, G.; Mazzeo, N.; et al. Warmer climates boost cyanobacterial dominance in shallow lakes. *Glob. Chang. Biol.* **2012**, *18*, 118–126. [CrossRef]

19.	Richardson, J.; Miller, C.; Maberly, S.C.; Taylor, P.; Globevnik, L.; Hunter, P.; Jeppesen, E.; Mischke, U.; Moe, S.J.; Pasztaleniec, A.; et al. Effects of multiple stressors on cyanobacteria abundance vary with lake type. *Glob. Chang. Biol.* **2018**, *24*, 5044–5055. [CrossRef]

20.	Jeppesen, E.; Mehner, T.; Winfield, I.J.; Kangur, K.; Sarvala, J.; Gerdeaux, D.; Rask, M.; Malmquist, H.J.; Holmgren, K.; Volta, P.; et al. Impacts of climate warming on the long-term dynamics of key fish species in 24 European lakes. *Hydrobiologia* **2012**, *694*, 1–39. [CrossRef]

21.	Crain, C.M.; Kroeker, K.; Halpern, B.S. Interactive and cumulative effects of multiple human stressors in marine systems. *Ecol. Lett.* **2008**, *11*, 1304–1315. [CrossRef]

22.	Piggott, J.J.; Townsend, C.R.; Matthaei, C.D. Reconceptualizing synergism and antagonism among multiple stressors. *Ecol. Evol.* **2015**, *5*, 1538–1547. [CrossRef] [PubMed]

23.	Folt, C.L.; Chen, C.Y.; Moore, M.V.; Burnaford, J. Synergism and antagonism among multiple stressors. *Limnol. Oceanogr.* **1999**, *44*, 864–877. [CrossRef]

24.	Rigosi, A.; Carey, C.C.; Ibelings, B.W.; Brookes, J.D. The interaction between climate warming and eutrophication to promote cyanobacteria is dependent on trophic state and varies among taxa. *Limnol. Oceanogr.* **2014**, *59*, 99–114. [CrossRef]

25.	Matthaei, C.D.; Piggott, J.J.; Townsend, C.R. Multiple stressors in agricultural streams: Interactions among sediment addition, nutrient enrichment and water abstraction: Sediment, nutrients & water abstraction. *J. Appl. Ecol.* **2010**, *47*, 639–649.

26.	Piggott, J.J.; Lange, K.; Townsend, C.R.; Matthaei, C.D. Multiple Stressors in Agricultural Streams: A Mesocosm Study of Interactions among Raised Water Temperature, Sediment Addition and Nutrient Enrichment. *PLoS ONE* **2012**, *7*, e49873. [CrossRef] [PubMed]

27.	Townsend, C.R.; Uhlmann, S.S.; Matthaei, C.D. Individual and combined responses of stream ecosystems to multiple stressors. *J. Appl. Ecol.* **2008**, *45*, 1810–1819. [CrossRef]

28. Miguet, P.; Logez, M.; Argillier, C. *Incertitudes Liées aux Interactions de Pressions. Effet des Interactions Entre les Pressions sur les Métriques Constitutives des Indicateurs IIL et IIR—Rapport AQUAREF. Personal Communication*; Irstea: Aix-en-provence, France, 2018.

29. Paine, R.T.; Tegner, M.J.; Johnson, E.A. Compounded Perturbations Yield Ecological Surprises. *Ecosystems* **1998**, *1*, 535–545. [CrossRef]

30. Villar-Argaiz, M.; Medina-Sánchez, J.M.; Biddanda, B.A.; Carrillo, P. Predominant Non-additive Effects of Multiple Stressors on Autotroph C:N:P Ratios Propagate in Freshwater and Marine Food Webs. *Front. Microbiol.* **2018**, *9*, 69. [CrossRef]

31. Lange, K.; Townsend, C.R.; Matthaei, C.D. Inconsistent Relationships of Primary Consumer N Stable Isotope Values to Gradients of Sheep/Beef Farming Intensity and Flow Reduction in Streams. *Water* **2019**, *11*, 2239. [CrossRef]

32. Walters, A.W.; Bartz, K.K.; Mcclure, M.M. Interactive Effects of Water Diversion and Climate Change for Juvenile Chinook Salmon in the Lemhi River Basin (U.S.A.): Water Diversion and Climate Change. *Conserv. Biol.* **2013**, *27*, 1179–1189. [CrossRef]

33. Wenger, S.J.; Isaak, D.J.; Luce, C.H.; Neville, H.M.; Fausch, K.D.; Dunham, J.B.; Dauwalter, D.C.; Young, M.K.; Elsner, M.M.; Rieman, B.E.; et al. Flow regime, temperature, and biotic interactions drive differential declines of trout species under climate change. *Proc. Natl. Acad. Sci. USA* **2011**, *108*, 14175–14180. [CrossRef]

34. Winder, M.; Reuter, J.E.; Schladow, S.G. Lake warming favours small-sized planktonic diatom species. *Proc. R. Soc. B Biol. Sci.* **2009**, *276*, 427–435. [CrossRef]

35. Jeppesen, E.; Sondergaard, M.; Jensen, J.P.; Havens, K.E.; Anneville, O.; Carvalho, L.; Coveney, M.F.; Deneke, R.; Dokulil, M.T.; Foy, B.; et al. Lake responses to reduced nutrient loading—An analysis of contemporary long-term data from 35 case studies. *Freshw. Biol.* **2005**, *50*, 1747–1771. [CrossRef]

36. Jeppesen, E.; Peder Jensen, J.; SØndergaard, M.; Lauridsen, T.; Landkildehus, F. Trophic structure, species richness and biodiversity in Danish lakes: Changes along a phosphorus gradient: A detailed study of Danish lakes along a phosphorus gradient. *Freshw. Biol.* **2000**, *45*, 201–218. [CrossRef]

37. Kratina, P.; Greig, H.S.; Thompson, P.L.; Carvalho-Pereira, T.S.A.; Shurin, J.B. Warming modifies trophic cascades and eutrophication in experimental freshwater communities. *Ecology* **2012**, *93*, 1421–1430. [CrossRef] [PubMed]

38. Bucak, T.; Saraoğlu, E.; Levi, E.E.; Nihan Tavşanoğlu, Ü.; İdİl Çakiroğlu, A.; Jeppesen, E.; Beklioğlu, M. The influence of water level on macrophyte growth and trophic interactions in eutrophic Mediterranean shallow lakes: A mesocosm experiment with and without fish: Effects of water level in warm shallow lakes. *Freshw. Biol.* **2012**, *57*, 1631–1642. [CrossRef]

39. Downing, J.A.; Plante, C.; Lalonde, S. Fish Production Correlated with Primary Productivity, not the Morphoedaphic Index. *Can. J. Fish. Aquat. Sci.* **1990**, *47*, 1929–1936. [CrossRef]

40. Kronvang, B.; Jeppesen, E.; Conley, D.J.; Søndergaard, M.; Larsen, S.E.; Ovesen, N.B.; Carstensen, J. Nutrient pressures and ecological responses to nutrient loading reductions in Danish streams, lakes and coastal waters. *J. Hydrol.* **2005**, *304*, 274–288. [CrossRef]

41. Meerhoff, M.; Teixeira-de Mello, F.; Kruk, C.; Alonso, C.; González-Bergonzoni, I.; Pacheco, J.P.; Lacerot, G.; Arim, M.; Beklioğlu, M.; Brucet, S.; et al. Environmental Warming in Shallow Lakes. In *Advances in Ecological Research*; Elsevier: London, UK, 2012; Volume 46, pp. 259–349.

42. Moss, B.; Stephen, D.; Balayla, D.M.; Becares, E.; Collings, S.E.; Fernandez-Alaez, C.; Fernandez-Alaez, M.; Ferriol, C.; Garcia, P.; Goma, J.; et al. Continental-scale patterns of nutrient and fish effects on shallow lakes: Synthesis of a pan-European mesocosm experiment. *Freshw. Biol.* **2004**, *49*, 1633–1649. [CrossRef]

43. Persson, L.; Diehl, S.; Johansson, L.; Andersson, G.; Hamrin, S.F. Shifts in fish communities along the productivity gradient of temperate lakes-patterns and the importance of size-structured interactions. *J. Fish Biol.* **1991**, *38*, 281–293. [CrossRef]

44. European Committee for Standardization (CEN). *Water Quality—Guidance on the Scope and Selection of Fish Sampling Methods (English Version EN 14757:2015)*; European Committee for Standardization; BSI Standards: Brussels, Belgium, 2015.

45. Prchalová, M.; Mrkvička, T.; Kubečka, J.; Peterka, J.; Čech, M.; Muška, M.; Kratochvíl, M.; Vašek, M. Fish activity as determined by gillnet catch: A comparison of two reservoirs of different turbidity. *Fish. Res.* **2010**, *102*, 291–296. [CrossRef]

46. Šmejkal, M.; Ricard, D.; Prchalová, M.; Říha, M.; Muška, M.; Blabolil, P.; Čech, M.; Vašek, M.; Jůza, T.; Monteoliva Herreras, A.; et al. Biomass and Abundance Biases in European Standard Gillnet Sampling. *PLoS ONE* **2015**, *10*, e0122437. [CrossRef] [PubMed]

47. Laplace-Treyture, C.; Rimet, F.; Anneville, O.; Druart, J.-C.; Barbe, J.; Dutartre, A. *Protocole Standardisé D'échantillonnage, de Conservation, D'observation et de Dénombrement du Phytoplancton en Plan D'eau Pour la Mise en Œuvre de la DCE: Version 3.3.1. Personal Communication*; Cemagref: Cestas, France, 2009; p. 44.

48. CEN-EN 15204. *Water Quality. Guidance Standard on the Enumeration of Phytoplankton Using Inverted Microscopy (Utermöhl Technique)*; AFNOR Normalisation: La plaine Saint-Denis, France, 2006; p. 39.

49. Derot, J.; Jamoneau, A.; Teichert, N.; Rosebery, J.; Morin, S.; Laplace-Treyture, C. Response of phytoplankton traits to environmental variables in French lakes: New perspectives for bioindication. *Ecol. Indic.* **2020**, *108*, 105659. [CrossRef]

50. Laplace-Treyture, C.; Hadoux, E.; Plaire, M.; Dubertrand, A.; Esmieu, P. *PHYTOBS v3.0: Outil de Comptage du Phytoplancton en Laboratoire et de Calcul de l'IPLAC. Version 3.0. Application JAVA. Personal Communication*; Irstea: Cestas, France, 2017.

51. Laplace-Treyture, C.; Feret, T. Performance of the Phytoplankton Index for Lakes (IPLAC): A multimetric phytoplankton index to assess the ecological status of water bodies in France. *Ecol. Indic.* **2016**, *69*, 686–698. [CrossRef]

52. CEN-EN 90-117. *Water Quality—Determination of Chlorophyll a and of a Pheopigments Index. Molecular Absorption Spectrometric Method*; AFNOR Normalisation: Paris, France, 1999; p. 11.

53. Bergström, L.; Karlsson, M.; Bergström, U.; Pihl, L.; Kraufvelin, P. Relative impacts of fishing and eutrophication on coastal fish assessed by comparing a no-take area with an environmental gradient. *Ambio* **2019**, *48*, 565–579. [CrossRef]

54. Van Dorst, R.M.; Gårdmark, A.; Svanbäck, R.; Beier, U.; Weyhenmeyer, G.A.; Huss, M. Warmer and browner waters decrease fish biomass production. *Glob. Chang. Biol.* **2019**, *25*, 1395–1408. [CrossRef]

55. Deceliere-Vergès, C.; Argillier, C.; Lanoiselée, C.; De Bortoli, J.; Guillard, J. Stability and precision of the fish metrics obtained using CEN multi-mesh gillnets in natural and artificial lakes in France. *Fish. Res.* **2009**, *99*, 17–25. [CrossRef]

56. Irz, P.; Odion, M.; Argillier, C.; Pont, D. Comparison between the fish communities of lakes, reservoirs and rivers: Can natural systems help define the ecological potential of reservoirs? *Aquat. Sci.* **2006**, *68*, 109–116. [CrossRef]

57. Daufresne, M.; Lengfellner, K.; Sommer, U. Global warming benefits the small in aquatic ecosystems. *Proc. Natl. Acad. Sci. USA* **2009**, *106*, 12788–12793. [CrossRef]

58. Ohlberger, J. Climate warming and ectotherm body size—From individual physiology to community ecology. *Funct. Ecol.* **2013**, *27*, 991–1001. [CrossRef]

59. White, E.P.; Ernest, S.K.M.; Kerkhoff, A.J.; Enquist, B.J. Relationships between body size and abundance in ecology. *Trends Ecol. Evol.* **2007**, *22*, 323–330. [CrossRef]

60. Guiet, J.; Poggiale, J.-C.; Maury, O. Modelling the community size-spectrum: Recent developments and new directions. *Ecol. Model.* **2016**, *337*, 4–14. [CrossRef]

61. Bartosiewicz, M.; Przytulska, A.; Deshpande, B.N.; Antoniades, D.; Cortes, A.; MacIntyre, S.; Lehmann, M.F.; Laurion, I. Effects of climate change and episodic heat events on cyanobacteria in a eutrophic polymictic lake. *Sci. Total Environ.* **2019**, *693*, 133414. [CrossRef] [PubMed]

62. Korkonen, S.; Weckström, J.; Korhola, A. Biogeography and ecology of freshwater chrysophyte cysts in Finland. *Hydrobiologia* **2019**, *847*, 497–499. [CrossRef]

63. Moss, B.; Kosten, S.; Meerhoff, M.; Battarbee, R.W.; Jeppesen, E.; Mazzeo, N.; Havens, K.; Lacerot, G.; Liu, Z.; De Meester, L.; et al. Allied attack: Climate change and eutrophication. *Inland Waters* **2011**, *1*, 101–105. [CrossRef]

64. Ignatiades, L. Redefinition of cell size classification of phytoplankton—A potential tool for improving the quality and assurance of data interpretation. *Mediterr. Mar. Sci.* **2015**, *17*, 56. [CrossRef]

65. Kruk, C.; Huszar, V.L.M.; Peeters, E.T.H.M.; Bonilla, S.; Costa, L.; LüRling, M.; Reynolds, C.S.; Scheffer, M. A morphological classification capturing functional variation in phytoplankton. *Freshw. Biol.* **2010**, *55*, 614–627. [CrossRef]

66. Stephen, D.; Balayla, D.M.; Becares, E.; Collings, S.E.; Fernandez-Alaez, C.; Fernandez-Alaez, M.; Ferriol, C.; Garcia, P.; Goma, J.; Gyllstrom, M.; et al. Continental-scale patterns of nutrient and fish effects on shallow lakes: Introduction to a pan-European mesocosm experiment. *Freshw. Biol.* **2004**, *49*, 1517–1524. [CrossRef]

67. Irz, P.; Michonneau, F.; Oberdorff, T.; Whittier, T.R.; Lamouroux, N.; Mouillot, D.; Argillier, C. Fish community comparisons along environmental gradients in lakes of France and north-east USA. *Glob. Ecol. Biogeogr.* **2007**, *16*, 350–366. [CrossRef]

68. Mehner, T.; Holmgren, K.; Lauridsen, T.L.; Jeppesen, E.; Diekmann, M. Lake depth and geographical position modify lake fish assemblages of the European 'Central Plains' ecoregion. *Freshw. Biol.* **2007**, *52*, 2285–2297. [CrossRef]

69. Danis, P.-A.; Ferrer, R.; Gevrey, M.; Argillier, C. *Seuils des Paramètres Physicochimiques Soutenant la Biologie—Plans d'eau Naturels—Rapport D'avancement. Personal Communication;* Irstea: Aix-en-Provence, France, 2012.

70. FAO. CEMAGREF. In *Crues et Apports: Manuel Pour L'estimation des Crues Décennales et des Apports Annuels Pour les Petits Bassins Versants non Jaugés de L'AFRIQUE Sahélienne et Tropicale Sèche;* Food and Agriculture Organization of the United Nations: Rome, Italy, 1996; pp. 61–77.

71. Livingstone, D.M.; Lotter, A.F. The relationship between air and water temperatures in lakes of the Swiss Plateau. *J. Paleolimnol.* **1998**, *19*, 181–198. [CrossRef]

72. Prats, J.; Danis, P.-A. An epilimnion and hypolimnion temperature model based on air temperature and lake characteristics. *Knowl. Manag. Aquat. Ecosyst.* **2019**, 8. [CrossRef]

73. Quintana-Seguí, P.; Le Moigne, P.; Durand, Y.; Martin, E.; Habets, F.; Baillon, M.; Canellas, C.; Franchisteguy, L.; Morel, S. Analysis of Near-Surface Atmospheric Variables: Validation of the SAFRAN Analysis over France. *J. Appl. Meteorol. Climatol.* **2008**, *47*, 92–107. [CrossRef]

74. Vidal, J.-P.; Martin, E.; Franchistéguy, L.; Baillon, M.; Soubeyroux, J.-M. A 50-year high-resolution atmospheric reanalysis over France with the Safran system. *Int. J. Climatol.* **2010**, *30*, 1627–1644. [CrossRef]

75. Vollenweider, R.A. Global problems of eutrophication and its control. *Symp. Biol. Hung* **1989**, *38*, 19–41.

76. European Union—SOeS Corine Land Cover. Available online: https://www.statistiques.developpement-durable.gouv.fr/corine-land-cover-0 (accessed on 2 December 2019).

77. Launois, L.; Veslot, J.; Irz, P.; Argillier, C. Selecting fish-based metrics responding to human pressures in French natural lakes and reservoirs: Towards the development of a fish-based index (FBI) for French lakes: Response of fish-based metrics to human pressures in French lakes. *Ecol. Freshw. Fish* **2011**, *20*, 120–132. [CrossRef]

78. Logez, M.; Pont, D.; Ferreira, M.T. Do Iberian and European fish faunas exhibit convergent functional structure along environmental gradients? *J. N. Am. Benthol. Soc.* **2010**, *29*, 1310–1323. [CrossRef]

79. Fox, J. Effect Displays in R for Generalised Linear Models. *J. Stat. Softw.* **2003**, *8*, 27. [CrossRef]

80. Elser, J.J.; Bracken, M.E.S.; Cleland, E.E.; Gruner, D.S.; Harpole, W.S.; Hillebrand, H.; Ngai, J.T.; Seabloom, E.W.; Shurin, J.B.; Smith, J.E. Global analysis of nitrogen and phosphorus limitation of primary producers in freshwater, marine and terrestrial ecosystems. *Ecol. Lett.* **2007**, *10*, 1135–1142. [CrossRef]

81. Rader, R.B.; Richardson, C.J. The effects of nutrient enrichment on algae and macroinvertebrates in the everglades: A review. *Wetlands* **1992**, *12*, 121–135. [CrossRef]

82. Smith, V.H.; Tilman, G.D.; Nekola, J.C. Eutrophication: Impacts of excess nutrient inputs on freshwater, marine, and terrestrial ecosystems. *Environ. Pollut.* **1999**, *100*, 179–196. [CrossRef]

83. Anderson, D.M.; Glibert, P.M.; Burkholder, J.M. Harmful algal blooms and eutrophication: Nutrient sources, composition, and consequences. *Estuaries* **2002**, *25*, 704–726. [CrossRef]

84. Jeppesen, E.; Kronvang, B.; Meerhoff, M.; Søndergaard, M.; Hansen, K.M.; Andersen, H.E.; Lauridsen, T.L.; Liboriussen, L.; Beklioglu, M.; Özen, A.; et al. Climate Change Effects on Runoff, Catchment Phosphorus Loading and Lake Ecological State, and Potential Adaptations. *J. Environ. Qual.* **2009**, *38*, 1930. [CrossRef] [PubMed]

85. Moss, B.; Mckee, D.; Atkinson, D.; Collings, S.E.; Eaton, J.W.; Gill, A.B.; Harvey, I.; Hatton, K.; Heyes, T.; Wilson, D. How important is climate? Effects of warming, nutrient addition and fish on phytoplankton in shallow lake microcosms: Climate change and phytoplankton. *J. Appl. Ecol.* **2003**, *40*, 782–792. [CrossRef]

86. Pörtner, H.-O.; Roberts, D.C.; Masson-Delmotte, V.; Zhai, P.; Tignor, M.; Poloczanska, E.; Mintenbeck, K.; Alegría, A.; Nicolai, M.; Okem, A.; et al. *IPCC Special Report on the Ocean and Cryosphere in a Changing Climate;* IPCC: Geneva, Switzerland, 2019.

87. Keller, W. Implications of climate warming for Boreal Shield lakes: A review and synthesis. *Environ. Rev.* **2007**, *15*, 99–112. [CrossRef]

88. Binzer, A.; Guill, C.; Rall, B.C.; Brose, U. Interactive effects of warming, eutrophication and size structure: Impacts on biodiversity and food-web structure. *Glob. Chang. Biol.* **2016**, *22*, 220–227. [CrossRef]

89. Downing, J.A.; Plante, C. Production of Fish Populations in Lakes. *Can. J. Fish. Aquat. Sci.* **1993**, *50*, 110–120. [CrossRef]

90. Atkinson, D. Temperature and Organism Size—A Biological Law for Ectotherms. In *Advances in Ecological Research*; Elsevier: London, UK, 1994; Volume 25, pp. 1–58.

91. Sentis, A.; Binzer, A.; Boukal, D.S. Temperature-size responses alter food chain persistence across environmental gradients. *Ecol. Lett.* **2017**, *20*, 852–862. [CrossRef]

92. Dodson, S.I.; Arnott, S.E.; Cottingham, K.L. The Relationship in Lake Communities between Primary Productivity and Species Richness. *Ecology* **2000**, *81*, 19. [CrossRef]

93. Bosse, K.R.; Sayers, M.J.; Shuchman, R.A.; Fahnenstiel, G.L.; Ruberg, S.A.; Fanslow, D.L.; Stuart, D.G.; Johengen, T.H.; Burtner, A.M. Spatial-temporal variability of in situ cyanobacteria vertical structure in Western Lake Erie: Implications for remote sensing observations. *J. Gt. Lakes Res.* **2019**, *45*, 480–489. [CrossRef]

94. Dalu, T.; Wasserman, R.J. Cyanobacteria dynamics in a small tropical reservoir: Understanding spatio-temporal variability and influence of environmental variables. *Sci. Total Environ.* **2018**, *643*, 835–841. [CrossRef]

95. Emmrich, M.; Pédron, S.; Brucet, S.; Winfield, I.J.; Jeppesen, E.; Volta, P.; Argillier, C.; Lauridsen, T.L.; Holmgren, K.; Hesthagen, T.; et al. Geographical patterns in the body-size structure of European lake fish assemblages along abiotic and biotic gradients. *J. Biogeogr.* **2014**, *41*, 2221–2233. [CrossRef]

96. Pickering, A.D. Growth and stress in fish production. *Aquaculture* **1993**, *111*, 51–63. [CrossRef]

97. O'Gorman, E.J.; Ólafsson, Ó.P.; Demars, B.O.L.; Friberg, N.; Guðbergsson, G.; Hannesdóttir, E.R.; Jackson, M.C.; Johansson, L.S.; McLaughlin, Ó.B.; Ólafsson, J.S.; et al. Temperature effects on fish production across a natural thermal gradient. *Glob. Chang. Biol.* **2016**, *22*, 3206–3220. [CrossRef] [PubMed]

98. Logez, M.; Pont, D. Global warming and potential shift in reference conditions: The case of functional fish-based metrics. *Hydrobiologia* **2013**, *704*, 417–436. [CrossRef]

99. Mondy, C.P.; Villeneuve, B.; Archaimbault, V.; Usseglio-Polatera, P. A new macroinvertebrate-based multimetric index (I2M2) to evaluate ecological quality of French wadeable streams fulfilling the WFD demands: A taxonomical and trait approach. *Ecol. Indic.* **2012**, *18*, 452–467. [CrossRef]

100. Argillier, C.; Caussé, S.; Gevrey, M.; Pédron, S.; De Bortoli, J.; Brucet, S.; Emmrich, M.; Jeppesen, E.; Lauridsen, T.; Mehner, T.; et al. Development of a fish-based index to assess the eutrophication status of European lakes. *Hydrobiologia* **2013**, *704*, 193–211. [CrossRef]

101. Blabolil, P.; Logez, M.; Ricard, D.; Prchalová, M.; Říha, M.; Sagouis, A.; Peterka, J.; Kubečka, J.; Argillier, C. An assessment of the ecological potential of Central and Western European reservoirs based on fish communities. *Fish. Res.* **2016**, *173*, 80–87. [CrossRef]

102. Teichert, N.; Lepage, M.; Sagouis, A.; Borja, A.; Chust, G.; Ferreira, M.T.; Pasquaud, S.; Schinegger, R.; Segurado, P.; Argillier, C. Functional redundancy and sensitivity of fish assemblages in European rivers, lakes and estuarine ecosystems. *Sci. Rep.* **2017**, *7*, 17611. [CrossRef]

103. Turner, B.L.; Kasperson, R.E.; Matson, P.A.; McCarthy, J.J.; Corell, R.W.; Christensen, L.; Eckley, N.; Kasperson, J.X.; Luers, A.; Martello, M.L.; et al. A framework for vulnerability analysis in sustainability science. *Proc. Natl. Acad. Sci. USA* **2003**, *100*, 8074–8079. [CrossRef]

104. Parravicini, V.; Villéger, S.; McClanahan, T.R.; Arias-González, J.E.; Bellwood, D.R.; Belmaker, J.; Chabanet, P.; Floeter, S.R.; Friedlander, A.M.; Guilhaumon, F.; et al. Global mismatch between species richness and vulnerability of reef fish assemblages. *Ecol. Lett.* **2014**, *17*, 1101–1110. [CrossRef]

105. Teichert, N.; Argillier, C.; Lepage, M.; Segouis, A.; Schinegger, R.; Palt, M.; Schmutz, S.; Segurado, P.; Ferreira, M.T.; Chust, G.; et al. *Deliverable D5.1-5: Reports on Stressor Classification and Effects at the European Scale: New Functional Diversity Indices Allowing Assessing Vulnerability in Abiotic Multi-Stressor Context*; MARS Project, Department of Aquatic Ecology/Faculty of Biology, Centre for Water and Environmental Research (ZWU): Duisburg-Essen, Germany, 2016; pp. 515–543.

106. Allan, J.D.; McIntyre, P.B.; Smith, S.D.P.; Halpern, B.S.; Boyer, G.L.; Buchsbaum, A.; Burton, G.A.; Campbell, L.M.; Chadderton, W.L.; Ciborowski, J.J.H.; et al. Joint analysis of stressors and ecosystem services to enhance restoration effectiveness. *Proc. Natl. Acad. Sci. USA* **2013**, *110*, 372–377. [CrossRef] [PubMed]
107. Brown, C.J.; Saunders, M.I.; Possingham, H.P.; Richardson, A.J. Managing for Interactions between Local and Global Stressors of Ecosystems. *PLoS ONE* **2013**, *8*, e65765. [CrossRef] [PubMed]

Article

Disentangling the Effects of Multiple Stressors on Large Rivers Using Benthic Invertebrates—A Study of Southeastern European Large Rivers with Implications for Management

Gorazd Urbanič [1,*], Zlatko Mihaljević [2], Vesna Petkovska [3] and Maja Pavlin Urbanič [1]

[1] URBANZERO Institute for holistic environmental management, Ltd., Selo pri Mirni 17, 8233 Mirna, Slovenia; maja@urbanzeroinstitute.com

[2] Faculty of Science, Department of Biology, University of Zagreb, Rooseveltov trg 6, 10000 Zagreb, Croatia; zlatko.mihaljevic@biol.pmf.hr

[3] Clinical Institute for Occupational, Traffic and Sports Medicine, University Medical Centre Ljubljana, Poljanski nasip 58, 1000 Ljubljana, Slovenia; v.petkovska@gmail.com

* Correspondence: gorazd@urbanzeroinstitute.com; Tel.: +386-41-509-933

Received: 6 January 2020; Accepted: 22 February 2020; Published: 25 February 2020

Abstract: Predicting anthropogenic actions resulting in undesirable changes in aquatic systems is crucial for the development of effective and sustainable water management strategies. Due to the co-occurrence of stressors and a lack of appropriate data, the effects on large rivers are difficult to elucidate. To overcome this problem, we developed a partial canonical correspondence analyses (pCCA) model using 292 benthic invertebrate taxa from 104 sites that incorporated the effects of three stressors groups: hydromorphology, land use, and water quality. The data covered an environmental gradient from near-natural to heavily altered sites in five large rivers in Southeastern Europe. Prior to developing the multi-stressor model, we assessed the importance of natural characteristics on individual stressor groups. Stressors proved to be the dominant factors in shaping benthic invertebrate assemblages. The pCCA among stressor-groups showed that unique effects dominated over joint effects. Thus, benthic invertebrate assemblages were suitable for disentangling the specific effect of each of the three stressor groups. While the effects of hydromorphology were dominant, both water quality and land use effects were nearly equally important. Quantifying the specific effects of hydromorphological alterations, water quality, and land use will allow water managers to better understand how large rivers have changed and to better define expectations for ecosystem conditions in the future.

Keywords: large rivers; multiple stressors; land use; hydromorphology; water quality; river-basin management; Southeastern Europe

---

1. Introduction

It is recognised that large rivers are economically important, but they also provide various ecosystem services and hence require sustainable management. The European Water Framework Directive [1] requires the identification of significant anthropogenic pressures and the assessment of their impacts on water bodies. Thus, we need to correctly predict human activities that create unacceptable impacts on large rivers. While the sources of stress in large rivers are numerous [2,3], little is known about the prevalence, spatial patterns, interactions with the natural environment and co-occurrence of stressors and their effects [4]. The effects of multiple stressors are difficult to predict due to the complexity of the interactions among stressors [5,6]. Thus, the effects of the individual stressor may be masked by the presence of other stressors.

Human pressures and land use patterns have long been recognised as a threat to the functioning and ecological integrity of aquatic ecosystems, as impacts on habitats, water quality, and biota involve complex pathways, e.g., [7,8]. High amounts of pollutants and nutrients have been discharged into large rivers as a result of industrial development, urbanization and intensive agriculture [8]. During the 19th and 20th centuries, stream regulation transformed large rivers to allow for navigation and power generation at the expense of habitat loss [9,10]. Large rivers became impounded, and their channels straightened and separated from oxbow lakes by levees to protect human settlements against floods [11–13]. These activities generally reduced longitudinal connectivity and connectivity between the main channel and adjacent floodplain channels [14], disturbing the natural gradients of chemical and physical parameters along large river courses were disturbed.

Aquatic communities are altered on a relatively predictable gradient from natural, e.g., undisturbed or minimally disturbed conditions, to severely altered conditions [15]. Ecological studies of large rivers are usually limited to individual rivers and are rarely based on data along the whole environmental gradient from near-natural to heavily altered sites. The reason might be that in individual large rivers, especially of developed countries including Europe, few near-natural remain [16]. However, in Southeastern Europe, despite large, altered stretches, the large rivers contain some of the last natural, free-flowing stretches in Europe. The Kupa and Una Rivers are in near natural conditions along their entire courses. The middle and lower stretches of the Drava and Mura Rivers are rare examples of unregulated, very large European rivers. Major lower sections of the Sava River still exhibit a relatively natural geomorphic structure and hydrological regime and are fringed by large protected wetlands.

Certain natural characteristics (e.g., catchment characteristics, depth, channel pattern) play a role in structuring benthic invertebrate communities, even in large river and at the regional scale e.g., [17]. Thus, the differences in these characteristics across large rivers must be accounted for before the impacts of stressors can be examined. Aside from natural conditions, hydromorphological alterations (the concept of 'hydromorphology' is a term introduced by the EC Water Framework Directive [1] that includes hydrological, morphological, and river continuity characteristics), land use, and water quality profoundly affect benthic invertebrates in rivers. Understanding the specific and joint effects of these stressors is of critical importance for developing effective river basin management plans to shape environmental policy.

In this study, we examined the unique and joint effects (two or more factors) of natural factors and major stressors (hydromorphology, land use, and water quality) on the invertebrate fauna of Southeastern European large rivers using the data along the entire environmental gradient from near-natural sites up to heavily altered sites. The term stressor(s) refers to variable(s) of anthropogenic landscape changes and local abiotic stream conditions that reflect human activities, and herein is used in this sense. Natural factors not influenced by anthropogenic disturbance are referred to using the term typology. We posed three general hypotheses regarding benthic invertebrate responses to natural factors (typology) and major stressors:

(1) Stressors and natural factors play a key role in structuring benthic invertebrate communities in the large rivers of a certain region (e.g., Southeastern Europe), thus differences in natural characteristics must be accounted for before the impacts of stressors can be isolated.

(2) Hydromorphology, land use, and water quality have distinct individual effects on structuring stream benthic macroinvertebrate assemblages.

(3) Specific stressor effects of hydromorphology, land use, and water quality are more important than their joint effects in structuring the benthic macroinvertebrate assemblages of large rivers and thus benthic invertebrates can be used to disentangle the effects of these stressors on large rivers.

2. Materials and Methods

2.1. Study Area

The study was conducted in an area of two neighbouring countries: Slovenia with a total area of 20,273 km^2 and 4573 km of rivers with catchments larger than 10 km^2, and Croatia with a total area of 56,594 km^2 and 12,884 km of rivers with catchments larger than 10 km^2 (Figure 1). The rivers in each of the countries belong either to the Danube or Adriatic River Basin, though this study included only rivers of the Danube River Basin. The Danube River Basin covers 16,381 km^2 (80.8%) of the Slovenian territory and 35,101 km^2 (62%) of Croatia. The landscape within this basin is diverse in altitude and slope and features different river section types [18]. This study was limited to five major rivers of the Danube River Basin: the Drava with Mura, and the Sava with its tributaries Kupa/Kolpa and Una (Table 1). The Sava and Drava Rivers are among the largest discharge tributaries of the Danube River (1st and 4th, respectively) and represent some of the best-preserved rivers in Europe in terms of their biological and landscape diversity. The Sava River springs in Slovenia as a gravel-bed river under Alpine influences, the channel in Slovenia changes from simple straight to braided, before gaining its meandering course downstream of Zagreb and continuing to its mouth in Belgrade (Serbia). The Sava River is considered by nature conservationists and scientists to be one of the crown jewels of European nature [19]. The Drava River crosses ecoregions from high Alpine mountains to the Pannonian-Illyrian plain and features all typical fluvio-morphological river types from straight to braided to meandering channels. The lower Drava with the lower Mura River constitutes a 380 km free-flowing and semi-natural watercourse and represents one of the last remaining continuous, riverine landscapes in Central Europe [18]. Only stretches with a catchment area from between 5000 and 64,000 km^2 and altitudes between 74 and 338 m were included in this analysis.

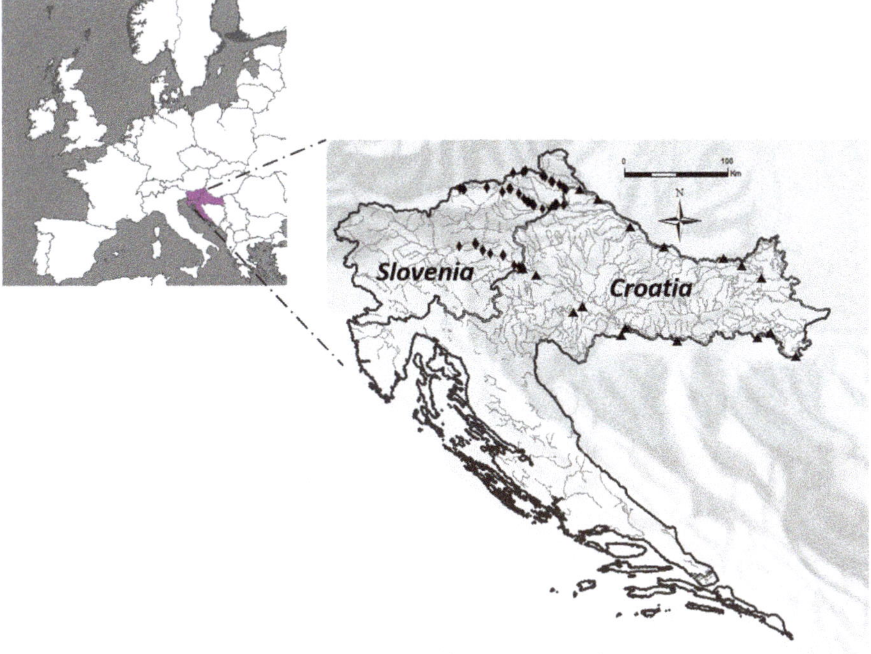

Figure 1. Study area showing the examined larger rivers and sampling sites (diamonds).

Table 1. Main characteristics of sampled rivers with the number of sampling sites and samples.

Country	River	Eco-Hydromorphological Type	Catchment Size Range (km^2)	Altitude Range (m a.s.l.)	No. Sites (Samples)
Slovenia	Drava	Intermountain	11,720–13,091	253–338	6 (14)
Slovenia	Drava	Lowland-braided	13,189–15,079	178–236	12 (23)
Slovenia	Mura	Lowland-braided	9784–10,506	165–246	11 (21)
Slovenia	Sava	Intermountain	4946–5203	191–222	3 (7)
Slovenia	Sava	Lowland-deep	7151–7655	154–191	4 (8)
Slovenia	Sava	Lowland-braided	7782–10,411	132–139	3 (9)
Croatia	Mura	Lowland-braided	10,930–10,930	153–153	1 (1)
Croatia	Mura	Lowland-deep	11,731–11,731	141–141	1 (1)
Croatia	Drava	Lowland-braided	14,363–31,038	122–190	2 (3)
Croatia	Drava	Lowland-deep	33,916–39,982	81–100	4 (4)
Croatia	Sava	Lowland-braided	10,997–12,316	113–132	2 (2)
Croatia	Sava	Lowland-deep	12,884–64,073	74–91	6 (7)
Croatia	Kupa	Lowland-deep	9184–9184	92–92	1 (2)
Croatia	Una	Lowland-deep	9368–9368	94–94	1 (2)
Total			4946–64,073	74–338	57 (104)

2.2. Environmental Variables

The sampling sites cover near-natural to highly disturbed conditions, reflecting the various disturbance levels caused by different stressors, e.g., hydromorphological alteration, catchment land-use, and water quality (Table 2). A total of 34 environmental variables were measured or calculated and classified into four groups: typology (natural), hydromorphology, land use, and water quality. The data for the five typology variables were obtained from the GIS database, the hydrological databases of the Slovenian Environment Agency (ARSO) and Croatian Meteorological and Hydrological Service (DHMZ), and from field analyses. Altitude and slope were calculated using a digital elevation model with 5 m accuracy. The natural predominant substrate was classified into three classes, as the smaller fraction (psammal-1), small to medium fraction (psammal/akal-2) and larger fraction (lithal-3), and the mean depth at low water level was defined as 1 when <1.5 m and 2 when >1.5 m.

Physical and chemical data were obtained monthly or at least four times a year (each season) from the national surface water monitoring programmes. In these analyses, only those 13 parameters were considered where data were available for all selected sites (Table 2): conductivity, pH, oxygen concentration, oxygen saturation, water temperature, COD($K_2Cr_2O_7$), BOD$_5$, orthophosphate, total nitrogen, ammonium, nitrite, nitrate, and total suspended solids. In the analyses, the median of data gathered for each parameter in the year of benthic invertebrate sampling was used.

Land use variables were defined from the share of land use categories at the catchment scale, extracted from Corine Land Cover (CLC) data [20] using ArcGIS version 10.2.1 (Esri Corp., Redlands, CA, USA). The categories were combined into five land use variables: urban land use (CLC class 1), natural and semi-natural land use (CLC classes 3, 4, 5), non-intensive agriculture land use (CLC categories 2.3.1, 2.4.3, 2.4.4), intensive agriculture land use on arable land (CLC categories 2.1), and intensive agriculture land use on non-arable land (CLC categories 2.2, 2.4.1, 2.4.2). Nine hydromorphological (HM) variables were selected; four discharge parameters and five HM indices of the Slovenian hydromorphological (SIHM) assessment method [17,21,22]. The SIHM method was applied to examine habitat quality, habitat modifications, and the influence of main upstream barriers/impoundments were considered. First, a river habitat survey [23,24] was performed once for each sampling site and the data was used to calculate the two morphological indices [21,22]: river habitat quality index (RHQ), and river habitat modification index (RHM). Normalised values (converted to a common scale of 0-1; RHQnor, RHMnor; [17]) were used. We first defined the eco-hydromorphology types of the considered river stretches according to [17] (Table 1, Figure 1). The RHM index was normalised using the same values for all river types; a reference value and a lower anchor of 0 and 112, respectively. RHQ values were normalised using type specific reference values.

For intermountain and lowland-deep eco-hydromorphology river types, a reference value of RHQ = 237 was used, whereas for lowland-braided the RHQ was set at 327. The lower anchor was the same (RHQ = 116) for all river types. The data on impoundments recorded in the catchment of each sampling site was used to calculate the hydrological modification index (HLM; [17,21]). Combining the indices RHQnor, RHMnor and HLM two HM indices were calculated: hydromorphological modification index (HMM) and hydromorphological quality and modification index (HQM) [17,21,22]. Hydrological variables were obtained from the available data on discharge from national monitoring gauging stations (ARSO, DHMZ). In addition to the mean daily value of discharge measured on the day of benthic invertebrate sampling (Q), the mean annual discharge (MQ), the lowest annual discharge (daily average; NQ), and the highest annual discharge (daily average; HQ) were also calculated for the sampling-year period.

2.3. Benthic Invertebrates

Biological data were obtained as part of the WFD monitoring and assessment system development programmes in Slovenia and Croatia between 2005 and 2011. In total, 104 samples were collected at 57 sites: 39 sites (82 samples) in Slovenia and 18 sites (22 samples) in Croatia (Figure 1, Appendix A). Some sites were sampled several times, but not more than once per year. Benthic invertebrates were collected during low to medium discharge using a multi-habitat sampling approach. Samples were collected in the wadeable part (up to 1.2 m) of the main channel or in the littoral zone of the impoundments to a depth of 1 m using a hand net (frame 25 × 25 cm, mesh-size: 500 μm). On each occasion, at every site, 20 sub-sampling units with a total sampling area of 1.25 m^2 were taken along a 100–250 m river stretch. The sampling procedure in Slovenia followed the standardized Slovenian river bioassessment protocol [17,25,26]. Twenty sampling units were selected in proportion to the coverage of the microhabitat types [17,24]. Microhabitat types were defined as the combination of substrate and flow type with at least 5% coverage. The channel substrate of each sampling site was classified according to [27], and flow characteristics according to [27,28]. Sampling units were pooled, preserved with 96% ethanol in the field and transferred to the lab for further processing. Each sample was sub-sampled, and the benthic organisms from a quarter of the whole field sample were identified and enumerated [29]. In Croatia, samples were collected according to the AQEM sampling strategy [27]. A total of 20 sampling units were sampled from representative substrates (i.e., substrates >5% coverage in the sample reach). At sampling sites with homogenous substratum (sand and other soft sediments) 10 sub-sampling units were taken instead of 20 (five sampling sites). In such cases, the sample was taken by pushing the hand net through the upper part (2–5 cm) of the substratum. The sampling units were pooled, preserved with 96% ethanol in the field, and transferred to the lab for further processing. In 2006, a more elaborate sub-sampling design was used, and habitat (substrate)-specific subsampling units were pooled and analysed as separate samples. In the lab, at least 1/6 of the sample was sorted until the minimum targeted number of 500 (habitat-specific samples) or 700 individuals (multi-habitat samples) was reached. Benthic invertebrates were identified usually to the species and genus level, though Oligochaeta and Diptera were identified to the (sub) family and genus level (Appendix B).

Table 2. Groups of environmental variables with their Median, Min (minimum) and Max (maximum) values.

Environmental Variable	Unit	Variable Group	Code	Median (Min-Max)	Transformation
Catchment size	km^2	Typology	C_size	11,720.1 (4945.8–64,073)	log(x + 1)
Depth mean	Classified 1–2	Typology	Depth	1 (1–2)	
Slope	(%)	Typology	slope	0.9 (0–3.6)	
Altitude	(m a.s.l.)	Typology	altitude	190 (74–338)	
Substrate	Classified 1–3	Typology	substratum	3 (1–3)	
Water temperature	°C	Water quality	T	12 (8.6–20.5)	log(x + 1)
pH		Water quality	pH	8 (7.7–8.3)	log(x + 1)
Conductivity	µS/cm	Water quality	cond	339.5 (262–517)	log(x + 1)
Oxygen concentration	mg O$_2$/L	Water quality	DO	9.5 (7.4–11.5)	log(x + 1)
Oxygen saturation	(%)	Water quality	DOsat	89 (79.5–104.2)	log(x + 1)
Total suspended solids	mg/L	Water quality	TSS	7 (2.4–56)	log(x + 1)
Chemical oxygen demand (K$_2$Cr$_2$O$_7$)	mg O$_2$/L	Water quality	COD	5.7 (2.5–19.6)	log(x + 1)
Biochemical oxygen demand (5 days)	mg O$_2$/L	Water quality	BOD5	1.2 (0.7–2.9)	log(x + 1)
Orthophosphate	mg P/L	Water quality	PO4	0 (0–0.3)	log(x + 1)
Total nitrogen	mg N/L	Water quality	Ntot	1.6 (0.7–2.6)	log(x + 1)
Ammonium	mg N/L	Water quality	NH4	0 (0–0.3)	log(x + 1)
Nitrite	mg N/L	Water quality	NO2	0 (0–0.1)	log(x + 1)
Nitrate	mg N/L	Water quality	NO3	1.4 (0.5–2)	log(x + 1)
Urban land use	(%)	Land use	C_urb	3.1 (0.9–5)	arcsin(sqrt x)
Natural and semi-natural land use	(%)	Land use	C_nat	70.8 (55–78.6)	arcsin(sqrt x)
Non-intensive agriculture land use	(%)	Land use	C_agrE	12.1 (10.1–24.9)	arcsin(sqrt x)
Intensive agriculture-tilled land use	(%)	Land use	C_agrI1	3.6 (0.7–14.7)	arcsin(sqrt x)
Intensive agriculture-non-tilled land use	(%)	Land use	C_agrI2	7.2 (4.5–20.9)	arcsin(sqrt x)
Discharge	m^3/s	Hydromorphology	Q	119.2 (5.2–824)	log(x + 1)
Mean annual discharge	m^3/s	Hydromorphology	NQ	114.7 (6.6–648)	log(x + 1)
Lowest annual discharge	m^3/s	Hydromorphology	MQ	216.7 (9–998)	log(x + 1)
Highest annual discharge	m^3/s	Hydromorphology	HQ	483.3 (31.1–1530)	log(x + 1)
River habitat quality index	Total score [1]	Hydromorphology	RHQ	218.6 (43.5–324.3)	
River habitat modification index	Total score [1]	Hydromorphology	RHM	28.6 (0–116)	
Hydrological modification index	Total score [1]	Hydromorphology	HLM	0.8 (0–1)	
Hydromorphological modification index	Total score [1]	Hydromorphology	HMM	0.7 (0–1)	
Hydromorphological quality and modification index	Total score [1]	Hydromorphology	HQM	0.7 (0–1)	

[1] Calculated score of individual features according to the SIHM method [20].

2.4. Data Analyses

Direct ordination techniques were carried out to analyse associations among environmental variables and between different groups of environmental variables and benthic invertebrate assemblages. These analyses were performed using Canoco 5 [30]. Benthic invertebrate data were transformed ($\ln(x + 1)$) prior to analysis. In addition, some environmental variables were transformed prior to the analyses to approximate the normal distribution [31] (Table 2). Catchment size, water quality variables, and hydrological variables were transformed using $\log(x + 1)$, whereas land use data (proportional data) were transformed using arcsin(sqrt x). Spearman rank correlation coefficients (R_{Sp}) were calculated between all pairs of environmental variables using SPSS Statistics version 21.0 [32]. The rationale was to identify associations among the analysed groups of variables and to compare them among different datasets. Since sampling season of benthic invertebrates differed among the samples, prior to performing the direct ordination analysis, the importance of the temporal variable represented by the sampling day in a year was tested. As the temporal variable explained only a low percentage in the variance of the benthic invertebrate dataset, in comparison to the environmental variables, it was not included in the further analyses.

To determine the compositional gradient length the invertebrate data were analysed using detrended correspondence analysis (DCA; [33]). Since these gradient lengths were greater than two standard deviations, we assumed unimodal species responses and, thus, canonical correspondence analysis (CCA; [34]) and partial canonical correspondence analysis were applied [35]. For the first overview of the relationship between the environmental variables and benthic invertebrate data, a CCA analysis with an automatic forward selection routine was applied to all environmental variables. This process specified the effects that each environmental variable added to the explained variance of the species data (marginal effects) and the remaining effect that each variable added to the model once when other variables had already been loaded (conditional effects) [34]. Significant variables were selected with forward selection routine, using the Monte Carlo permutation test with 999 unrestricted permutations. The same procedure was then applied within variable groups (i.e., typology, hydromorphology, water quality, land use). The selected variables were used for partitioning the explained variance among benthic invertebrate assemblages using partial CCA (pCCA). This test allows for the investigation of the effects of one variable group, while eliminating the effects of other variable groups, and hence the partitioning of the variance into unique and joined effects of variable groups. The total explained variance among benthic invertebrate assemblages with forward selected environmental variables from three groups was partitioned into (i) the variance uniquely explained by each variable group, (ii) the variance explained by combined effects of each pair of variable groups, and (iii) the variance explained by combined effects of all three variable groups together.

3. Results

3.1. Relationships between the Variables

Spearman rank correlation (R_{Sp}) resulted in several statistically significant relationships ($P < 0.05$) between pairs of environmental variables (Appendices C–F). Strong correlations ($|R_{Sp}| > 0.70$) among the variables of different stressor-groups were rare; natural and semi-natural land use related positively to altitude and negatively to conductivity, whereas alternatively non-intensive agriculture land use and intensive agriculture-non-tilled land use were negatively correlated with altitude and positively with conductivity. Other strong correlations were observed within all the stressor-groups, with the exception of the typology group. In the hydromorphology group, strong positive correlations were observed among hydrological variables and among indices (HLM, HMM, HQM). Several variable pairs of different stressor-groups showed moderate correlations ($0.50 < |R_{Sp}| < 0.70$). The lowest number of moderate and strong correlations was observed between the groups hydromorphology and water quality or land use. However, the indices HLM, HMM, and HQM showed a moderate positive correlation with conductivity and a negative correlation with natural and semi-natural land use. In the typology group, hydrological indices showed a moderate positive correlation with mean depth and catchment size and negative with slope. Most of the pairwise correlations were weak ($|R_{Sp}| < 0.50$) or insignificant.

3.2. Benthic Invertebrate Response to Environmental Variables

The total amount of variance (inertia) in the species data was 5.808, including the 104 sites and 292 benthic invertebrate taxa (Appendix B). The total explained variance in the dataset, including all 32 environmental variables, was 2.366 (41%). When tested individually, the highest explanatory power was observed for conductivity (0.22) (Table 3). Additionally, each of the 11 other variables showed more than 50% explanation power (>0.11) of the best explanatory variable. The variables of all four groups showed considerable explanation power. The hydromorphology group was represented with four indices (HLM index, HDM index, RHQ index and HMM index). An additional four eutrophication variables represented the water quality group (nitrogen-total, nitrate, nitrite, orthophosphate). Typology was represented by depth and altitude, whereas land use by natural and semi-natural land use. The remaining other 21 variables exhibited a weaker explanation power.

Table 3. Percentage of benthic invertebrate assemblages variance explained by each environmental variable's independent (before forward selection) and conditional effects (after forward selection within each environmental variable group).

Environmental Variable	Variable Group	Before FS λ	After FS Groups λ	P	F
Depth	Typology	0.16	0.16	0.001	2.86
Altitude	Typology	0.14	0.14	0.001	2.51
Slope	Typology	0.11	0.07	0.122	1.3
Catchment size	Typology	0.1	0.09	0.001	1.73
Substratum	Typology	0.07	0.09	0.013	1.63
Conductivity	Water quality	0.22	0.22	0.001	3.93
Nitrogen-total	Water quality	0.14	0.06	0.444	1.01
Nitrate	Water quality	0.13	0.05	0.358	1.05
Nitrite	Water quality	0.13	0.06	0.273	1.09
Orthophosphate	Water quality	0.13	0.09	0.001	1.75
Ammonia	Water quality	0.11	0.09	0.011	1.68
COD	Water quality	0.1	0.07	0.066	1.29
Temperature	Water quality	0.08	0.07	0.04	1.34
BOD$_5$	Water quality	0.08	0.06	0.12	1.21
Dissolved oxygen saturation	Water quality	0.08	0.08	0.031	1.41
Dissolved oxygen concentration	Water quality	0.07	0.07	0.1	1.2
pH	Water quality	0.07	0.05	0.412	1.02
Total suspended solids	Water quality	0.06	0.07	0.033	1.36
Natural and semi-natural land use	Land use	0.12	0.12	<0.0001	2.19
Intensive agriculture-non-tilled land use	Land use	0.11	0.12	<0.0001	2.12
Intensive agriculture-tilled land use	Land use	0.11	0.06	0.187	1.16
Non-intensive agriculture land use	Land use	0.11	0.09	0.005	1.72
Urban land use	Land use	0.08	0.1	0.002	1.81
Hydrological modification index	Hydromorphology	0.18	0.18	0.001	3.29
Hydromorphological quality and modification index	Hydromorphology	0.17	0.08	0.011	1.62
River habitat quality index	Hydromorphology	0.15	0.08	0.001	1.61
Hydromorphological modification index	Hydromorphology	0.15	0.08	0.015	1.45
Discharge	Hydromorphology	0.1	0.08	0.015	1.51
River habitat modification index	Hydromorphology	0.1	0.1	0.002	1.73
Highest annual discharge	Hydromorphology	0.08	0.07	0.039	1.29
Lowest annual discharge	Hydromorphology	0.08	0.08	0.015	1.51
Mean annual discharge	Hydromorphology	0.08	0.05	0.528	0.96

Testing each explanatory group individually, 22 of 32 environmental variables significantly contributed to the explained variance (Table 3). Each variable group comprised four to eight forward selected variables, used in the variance partitioning. The highest number of selected variables was in the hydromorphology group (eight out of nine), followed by the water quality group (six out of 13), typology group (four out of five) and land use group (four out of five). In the hydromorphology group, a combination of alteration indices and hydrological conditions was observed. For the water quality group, the selected variables reflected eutrophication, organic pollution and some other human activities. In the typology group, a combination of catchment conditions (catchment size, altitude) and instream conditions (depth, substrate) was observed. The land use group reflected urbanisation, agriculture, and other non-natural land use.

3.3. Variance Partitioning between Typology and Stressor-Groups

Variance partitioning between the typology group and an individual stressor-group revealed the unique effects of the stressor-groups, explaining from 36% (land use) to 53% (hydromorphology) of the benthic invertebrate assemblages explained variability (Figure 2). The joint effects (% of the total explained variance) of each stressor group and the typology group were relatively small (8–20%) in comparison to pure stressor effects. Joint effects always represented <30% of the stressor-group total explained variability.

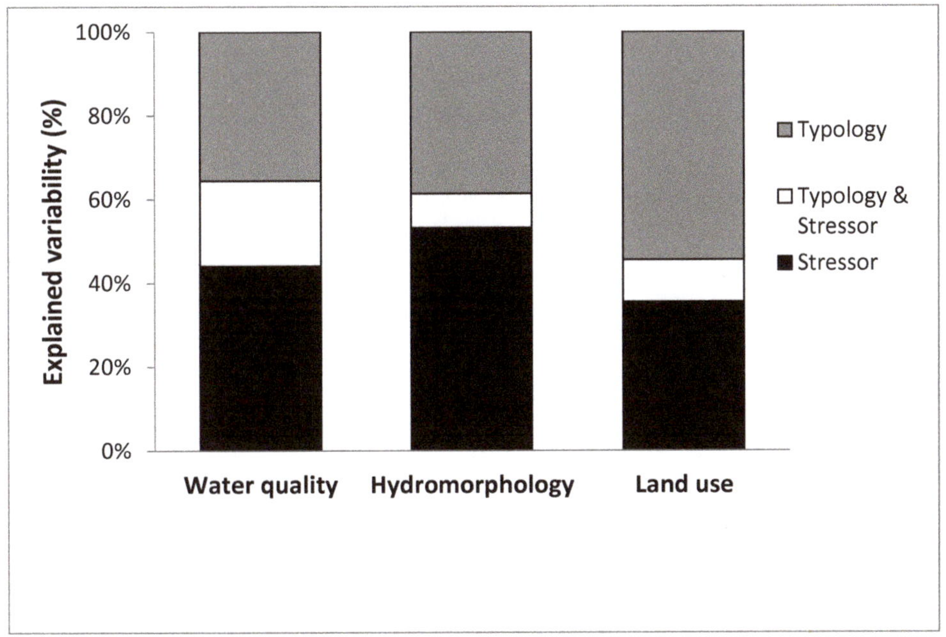

Figure 2. Unique and joint effect contribution of the typology group and each of the stressor variable groups (hydromorphology, land use, and water quality) to the explained variability of benthic invertebrate assemblages; given as percentage of joint and both unique contributions.

3.4. Variance Partitioning of Three Stressor Variable Groups of Environmental Variables

Variance partitioning was run with 18 variables, after a forward selection routine for each variable group separately. Clearly, the unique effects of variable groups were more important in explaining the variation in the benthic invertebrate composition than joint effects (84% and 16% of the explained variance, respectively, Figure 3). The highest share (36%) was explained by the hydromorphology group, followed by water quality (27%) and land use group (21%). The explanatory power of any joint effect was much smaller where interaction between water quality and hydromorphology groups was most important, accounting for 10%. Other interactions between group pairs were less important (≤4%) and the joint effects of all three variable groups explained only 2% of the variation.

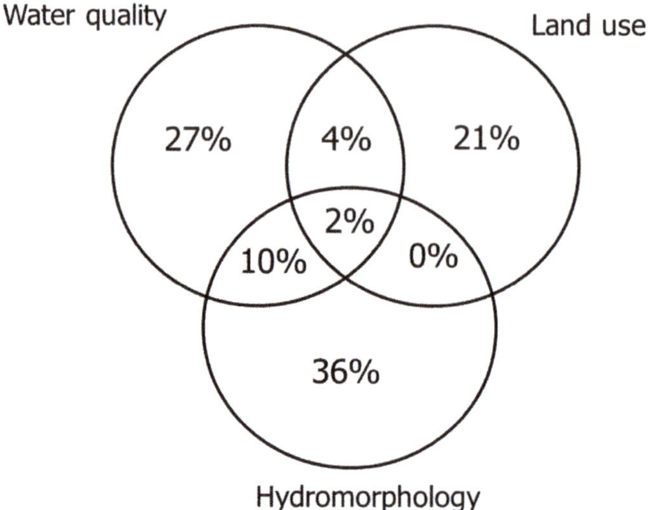

Figure 3. Venn diagram representing stressor-specific and joint effect contributions of the stressor variable groups to the explained variability of the benthic invertebrate assemblages. Explained variability is given as percentage of the total explained variance.

4. Discussion

Centuries of human activities including water pollution and habitat alterations have profoundly altered most large rivers and their aquatic assemblages. Conversion of native forests to agricultural and urban uses has increased concentrations of pollutants (e.g., nutrients), as well as habitat changes [36]. Only a few large European rivers still have stretches that appear to remain in their natural conditions. This study examined some of these stretches in Southeastern Europe, including natural and degraded large river stretches enabled us to cover the whole environmental gradient from near-natural up to heavily altered sites. Thus, our sampling sites exhibit a wide range of chemical and physical factors reflecting differences in habitat characteristics, land use and water quality.

Large rivers are unique ecosystems. Although they share some abiotic and biotic commonalities, certain natural characteristics play a role in structuring benthic invertebrate communities of large rivers, even within the same region (e.g., [17]). Bonada et al. [37] stated that isolating the natural variability along a large river course from the influence of water pollution, land use, and hydromorphological (HM) alterations is difficult due to their confounding effects. This study confirmed the presence of certain joint effects of natural characteristics and individual stressors, though the joint effects were found to be less conspicuous than the specific stressor effects. Our results indicated that regional data of large rivers can be pooled and stressor effects isolated, partly supporting the hypothesis that differences in natural characteristics must be accounted before the impacts of stressors can be determined. We found that, for large rivers, the joint effects also depend on the stressor group. Water quality showed the highest joint effects with typology reflecting that water pollution impacts depend on the natural characteristics of the large river. For example, the effects of nutrients are more evident in large rivers with slower water flow and higher water temperature [38,39]. Hydromorphological alterations showed the lowest joint effects with typology, which might reflect that HM alterations similarly change benthic habitats and communities of large rivers in way in all large rivers. This supports the findings of minimal differences in responses of benthic invertebrate assemblages to

HM alterations among large river types [17]. Land use showed an intermediate joint effect, possibly reflecting the combination of effects on water quality (e.g., eutrophication) and habitat characteristics (e.g., sedimentation) [25,36]. Due to the presence of multiple stressors and the lack of appropriate data, effects on large rivers can be difficult to elucidate [17,40]. In this study, pure stressor-specific effects on benthic invertebrates were heavily dominant over the joint effects, and hence, benthic invertebrate assemblages be useful in disentangling the effects of hydromorphology, water quality, and land use. Nevertheless, substantial joint effects were observed between HM alterations and water quality. The combination of both these stressors likely exerts substantial change in benthic invertebrate communities in large rivers. It is often observed that HM alterations (e.g., water abstraction, damming) also lead to water quality issues (e.g., eutrophication) [39,41,42]. The joint effects of the other two stressor groups and all three stressor groups together were small. It is known that land use changes impact water quality and HM conditions, thus, substantial joint effects could be expected. However, it seems that in Southeastern Europe, land use is not so intensive to severely influence water quality and/or HM conditions of large rivers. Moreover, river damming and channelling are key HM pressures impacting large river benthic habitats in the region [43,44]. We found that the effects of HM alterations were more significant than those of water quality and land use changes. Although the sequence and timing of individual stressor effects were not determined in this study, there is evidence that water pollution was most important during the early to mid-20th century [45,46]. Changes in land use have long been present but are intensifying, whereas HM alterations have become more significant in recent decades. Physical pressures have been identified as major causes for a potential failure of water bodies to meet the Water Framework Directive environmental objectives [4,47]. Nevertheless, this study showed that special attention should also be given to the effects of water pollution and land use.

Modelling the relationship between biological communities and environmental parameters has played an increasingly important role in ecology [48]. Such a predictive approach can lead to a better understanding of how the species composition can potentially be affected by human pressures, and is especially promising for use in conservation planning and resource management [49]. Partial canonical correspondence analyses (pCCA), proved to be a useful method for disentangling the effects of addressed stressors. However, there are limitations since the results also depend on selected variables, length of the gradient, and correlation among variables (e.g., [25,50,51]). In this study, the stressor variables were selected according to the reported impacts on large river aquatic communities [8,17,41]. In the water quality group, several other parameters could have been selected, though we chose the most relevant parameters from the eutrophication and organic pollution group. Strong correlations were found between environmental variables, but only within the stressor group. Therefore, it was possible to isolate the effects of different stressors on benthic invertebrates. The long environmental gradient of sites from near-natural conditions up to heavily altered sites is crucial for building a reliable model. It is also important to view the large river community dynamics not only in the context of environmental variables, but also in biotic interactions [52]. Alien species in particular might influence benthic invertebrate community responses (e.g., [50]), though this was not an issue in this study, as recorded alien species (e.g., *Corbicula fluminea*, *Dreissena polymorpha*, *Dikerogammarus villosus*, *D. haemobaphes*, *Jaera istri*) usually represented less than 5% of the benthic invertebrate assemblages' sample composition.

Understanding the impact of water pollution, hydromorphology, and land use change on the ecological status and ecosystem services is essential for developing effective river basin management plans (RBMPs) and shaping future environmental policy. Setting appropriate measures will enable environmental objectives to be achieved (e.g., good ecological status according to Water Framework Directive [1]). Relationships have previously been defined between the biota and water quality [53–55] what resulted in active river management for water quality improvement [56]. We showed that water quality issues still exist in large rivers and their effects also interact with HM alterations and land use. HM alterations are the dominant stressor in rivers throughout Europe [4,47], and many studies consider only HM alterations (e.g., [17,22,57–60]). We showed that in addition to HM alterations and water quality, land use impacts on benthic invertebrates are substantial. For example, increased urbanisation and intensive agriculture severely impact benthic invertebrate assemblages and the integrity of large rivers. Therefore, all major stressors need to be addressed and their effects disentangled to ensure implementation of sustainable river basin management strategies (e.g., Water Framework Directive). The integration of environmental objectives in sectoral policies (e.g., Common Agricultural Policy, Floods Directive, renewable energy, Natura 2000) having direct or indirect impacts on rivers and their catchments might help to achieve environmental objectives (e.g., good ecological status). Since most large European rivers have catchments that cross international borders, cooperation among countries is critical in planning and implementing management strategies. However, differences in development level, public opinion, and historical and political constraints can hinder attempts to achieve these common environmental objectives. Southeastern Europe is facing a range of development challenges, including the planning of new hydroelectric power plants, ongoing intensive urbanisation, and intensifying agriculture [13,25,44].

Public understanding of the importance of water quality, habitat conditions and land use in structuring aquatic assemblages in large rivers could provide a basis for greater support of effective large river protection and sustainable management efforts. However, the management agencies of Southeastern Europe need to change their paradigm of river water quality to the ecological quality of the river ecosystem, thereby supporting activities that would prevent large river deterioration as was observed in many parts of the world.

5. Conclusions

- We disentangled the specific effects of hydromorphology, water quality, and land use using benthic invertebrate assemblages.
- Joint effects of stressors and natural factors on benthic invertebrate assemblages depend on the stressor group.
- Stressors proved to be the dominant factors in shaping benthic invertebrate assemblages of Southeastern Europe large rivers. Effects of hydromorphology dominated over water quality and land use effects, though these were still substantial. Thus, all major stressors need to be addressed and their effects determined for the implementation of the sustainable river basin management strategies.
- Management agencies in Southeastern Europe need to change their paradigm from river water quality to the ecological quality of the river ecosystem, thereby supporting activities that will prevent large river deterioration.

Author Contributions: Conceptualization, G.U., Z.M., V.P. and M.P.U.; Formal analysis, G.U., V.P. and M.P.U.; Funding acquisition, G.U. and Z.M.; Investigation, G.U., Z.M. and V.P.; Methodology, G.U., V.P. and M.P.U.; Visualization, G.U. and V.P.; Writing-original draft, G.U., Z.M., V.P. and M.P.U.; Writing-Review & editing, G.U., Z.M., V.P. and M.P.U. All authors have read and agreed to the published version of the manuscript.

Funding: This research was conducted within the Slovenian-Croatian bilateral project "Benthic invertebrate based ecological status assessment of large rivers with management goals focused on hydromorphological alterations" (to G.U. and Z.M.). The study was also supported by the Ministry of the Environment and Spatial Planning of the Republic of Slovenia as a part of the National Programme for the Implementation of the EU Water Framework Directive, conducted at the Institute for Water of the Republic of Slovenia.

Acknowledgments: Authors gratefully acknowledge the project team members for their assistance both in the field and in the laboratory. We thank the Croatian Meteorological and Hydrological Service and Croatian Waters and Slovenian Environment Agency for providing hydrological and water physico-chemical data. We thank Maja Kerovec for assisting with GIS analysis.

Conflicts of Interest: The authors declare no conflict of interest.

Appendix A

Table A1. List of Sampling sites and number of collected samples.

River	Site	No. Samples	Latitude	Longitude
Drava	Belišće	1	45.6924	18.4187
Drava	Borl	1	46.3687	15.9903
Drava	Botovo	2	46.2592	16.9273
Drava	Bresternica	1	46.5678	15.5971
Drava	Brezno	2	46.5949	15.3154
Drava	Donji Miholjac	1	45.7831	18.2070
Drava	Dravograd	3	46.5884	15.0251
Drava	Frankovci	2	46.3974	16.1687
Drava	Grabe	2	46.3919	16.2542
Drava	Krčevina pri Ptuju	3	46.4403	15.8333
Drava	Križovljan Grad	1	46.3846	16.1157
Drava	Mariborski otok	4	46.5677	15.6137
Drava	Markovci	2	46.4106	15.8891
Drava	Ormož	3	46.3863	16.1206
Drava	Ptuj	2	46.4178	15.8690
Drava	Pušenci	2	46.4021	16.1571
Drava	Ranca	1	46.4108	15.8883
Drava	Ruše	3	46.5458	15.5083
Drava	Slovenja vas	1	46.4441	15.8130
Drava	Starše	1	46.4754	15.7702
Drava	Terezino Polje	1	45.9425	17.4822
Drava	Tribej	2	46.6020	14.9783
Drava	Višnjevac	1	45.5762	18.6452
Drava	Zgornji Duplek	2	46.5176	15.7143
Kupa	Brest	2	45.4424	16.2429
Mura	Bunčani	1	46.5985	16.1484
Mura	Ceršak	5	46.7062	15.6665
Mura	Gibina-Brod	2	46.5236	16.3391
Mura	Goričan	1	46.4154	16.7029
Mura	Gornja Bistrica	1	46.5404	16.2714
Mura	Konjišče	2	46.7193	15.8206
Mura	Mali Bakovci	1	46.6074	16.1280
Mura	Mele	2	46.6495	16.0504
Mura	Melinci	1	46.5719	16.2227
Mura	Mota	4	46.5504	16.2424
Mura	Peklenica	1	46.5105	16.4753
Mura	Petanjci	1	46.6492	16.0504
Mura	Trate	1	46.7070	15.7855
Sava	Boštanj	1	46.0110	15.2926
Sava	Brestanica	2	46.9873	15.4657
Sava	Brežice	1	45.8981	15.5903
Sava	Davor	2	45.1088	17.5247
Sava	Dolenji Leskovec	1	45.9860	15.4516
Sava	Drenje	1	45.8620	15.6924
Sava	Galdovo	1	45.4833	16.3935
Sava	Jasenovac	1	45.2633	16.8998
Sava	Jesenice na Dolenjskem	6	45.8609	15.6921
Sava	Mošenik	1	46.0922	14.9228
Sava	Podgračeno	2	45.8759	15.6500
Sava	Podkraj	3	46.1115	15.1158
Sava	Račinovci	1	44.8501	18.9661
Sava	Slavonski Šamac	1	45.0582	18.5093
Sava	Suhadol	3	46.1057	15.1253
Sava	Vrhovo	4	46.0445	15.2089
Sava	Zagreb-Jankomir	1	45.7911	15.8526
Sava	Županja	1	45.0685	18.6745
Una	Hrvatska Dubica	3	45.1900	16.7894

Appendix B

Table A2. List of the 292 benthic invertebrate taxa recorded at 104 river sampling sites. Ad.-adults, Lv.-larvae, Gr.-group.

Higher Taxon	Taxon
Turbellaria	*Dendrocoelum album*
Turbellaria	*Dendrocoelum lacteum*
Turbellaria	*Dugesia gonocephala*
Turbellaria	*Dugesia lugubris/polychroa*
Turbellaria	*Dugesia lugubris*
Turbellaria	*Dugesia tigrina*
Turbellaria	*Phagocata* sp.
Turbellaria	*Planaria torva*
Turbellaria	*Polycelis nigra/tenuis*
Nematoda	Nematoda Gen. sp.
Oligochaeta	Enchytraeidae Gen. sp.
Oligochaeta	*Haplotaxis gordioides*
Oligochaeta	*Eiseniella tetraedra*
Oligochaeta	Lumbriculidae Gen. sp.
Oligochaeta	*Lumbriculus variegatus*
Oligochaeta	*Rhynchelmis* sp.
Oligochaeta	*Stylodrilus heringianus*
Oligochaeta	*Stylodrilus* sp.
Oligochaeta	*Chaetogaster* sp.
Oligochaeta	*Dero* sp.
Oligochaeta	*Nais* sp.
Oligochaeta	*Ophidonais serpentina*
Oligochaeta	*Pristina* sp.
Oligochaeta	*Stylaria lacustris*
Oligochaeta	*Uncinais uncinata*
Oligochaeta	*Vejdovskiella comata*
Oligochaeta	*Vejdovskiella* sp.
Oligochaeta	*Propappus volki*
Oligochaeta	*Aulodrilus pluriseta*
Oligochaeta	*Branchiura sowerbyi*
Oligochaeta	*Peloscolex* sp.
Oligochaeta	*Peloscolex velutina*
Oligochaeta	Tubificidae juv without setae
Oligochaeta	Tubificidae juv with setae
Hirudinea	*Dina punctata*
Hirudinea	*Erpobdella nigricollis*
Hirudinea	*Erpobdella octoculata*
Hirudinea	*Erpobdella* sp.
Hirudinea	*Erpobdella testacea*
Hirudinea	*Erpobdella vilnensis*
Hirudinea	*Trocheta bykowskii*
Hirudinea	*Alboglossiphonia heteroclita*
Hirudinea	*Glossiphonia complanata*
Hirudinea	*Glossiphonia concolor*
Hirudinea	*Glossiphonia nebulosa*
Hirudinea	*Glossiphonia paludosa*
Hirudinea	*Glossiphonia* sp.
Hirudinea	*Glossiphonia verrucata*
Hirudinea	*Helobdella stagnalis*
Hirudinea	*Hemiclepsis marginata*
Hirudinea	*Theromyzon tessulatum*
Hirudinea	*Haemopis sanguisuga*

Table A2. *Cont.*

Higher Taxon	Taxon
Hirudinea	*Piscicola geometra*
Hirudinea	*Piscicola haranti*
Gastropoda	*Acroloxus lacustris*
Gastropoda	*Ancylus fluviatilis*
Gastropoda	*Bithynia tentaculata*
Gastropoda	*Bithynia* sp.
Gastropoda	*Borysthenia naticina*
Gastropoda	*Lithoglyphus naticoides*
Gastropoda	*Potamopyrgus antipodarum*
Gastropoda	*Sadleriana* sp.
Gastropoda	*Radix auricularia*
Gastropoda	*Radix balthica/labiata*
Gastropoda	*Radix balthica*
Gastropoda	*Radix labiata*
Gastropoda	*Esperiana acicularis*
Gastropoda	*Esperiana esperi*
Gastropoda	*Holandriana holandrii*
Gastropoda	*Theodoxus danubialis*
Gastropoda	*Theodoxus transversalis*
Gastropoda	*Physa fontinalis*
Gastropoda	*Physella acuta*
Gastropoda	*Gyraulus albus*
Gastropoda	*Gyraulus crista*
Gastropoda	*Planorbis carinatus*
Gastropoda	*Valvata cristata*
Gastropoda	*Valvata piscinalis*
Gastropoda	*Viviparus viviparus*
Bivalvia	*Dreissena polymorpha*
Bivalvia	*Musculium lacustre*
Bivalvia	*Pisidium* sp.
Bivalvia	*Sphaerium corneum*
Bivalvia	*Sphaerium* sp.
Bivalvia	*Sinanodonta woodiana*
Bivalvia	*Unio crassus*
Bivalvia	*Unio pictorum*
Bivalvia	*Unio tumidus*
Bivalvia	*Corbicula fluminea*
Arachnida	Hydrachnidia Gen. sp.
Amphipoda	*Synurella ambulans*
Amphipoda	*Gammarus fossarum*
Amphipoda	*Gammarus roeselii*
Amphipoda	*Corophium curvispinum*
Amphipoda	*Dikerogammarus haemobaphes*
Amphipoda	*Dikerogammarus villosus*
Amphipoda	*Niphargus* sp.
Isopoda	*Asellus aquaticus*
Isopoda	*Jaera istri*
Ephemeroptera	*Baetis buceratus*
Ephemeroptera	*Nigrobaetis digitatus*
Ephemeroptera	*Baetis fuscatus*
Ephemeroptera	*Baetis fuscatus/scambus*
Ephemeroptera	*Baetis liebenauae*
Ephemeroptera	*Baetis lutheri*
Ephemeroptera	*Baetis rhodani*
Ephemeroptera	*Baetis scambus*
Ephemeroptera	*Baetis* sp.
Ephemeroptera	*Baetis vardarensis*

Table A2. *Cont.*

Higher Taxon	Taxon
Ephemeroptera	*Baetis vernus*
Ephemeroptera	*Baetis buceratus/vernus*
Ephemeroptera	*Centroptilum luteolum*
Ephemeroptera	*Centroptilum* sp.
Ephemeroptera	*Cloeon dipterum*
Ephemeroptera	*Caenis* sp.
Ephemeroptera	*Brachycercus* sp.
Ephemeroptera	*Serratella ignita*
Ephemeroptera	*Ephemerella notata*
Ephemeroptera	*Ephemerella mucronata*
Ephemeroptera	*Torleya major*
Ephemeroptera	*Ephemera danica*
Ephemeroptera	*Ephemera* sp.
Ephemeroptera	*Ecdyonurus* sp.
Ephemeroptera	*Epeorus sylvicola*
Ephemeroptera	*Heptagenia* sp.
Ephemeroptera	*Heptagenia sulphurea*
Ephemeroptera	*Rhithrogena* sp.
Ephemeroptera	*Habroleptoides confusa*
Ephemeroptera	*Habrophlebia fusca*
Ephemeroptera	*Paraleptophlebia submarginata*
Ephemeroptera	*Oligoneuriella rhenana*
Ephemeroptera	*Potamanthus luteus*
Ephemeroptera	*Siphlonurus aestivalis*
Ephemeroptera	*Siphlonurus lacustris*
Ephemeroptera	*Siphlonurus* sp.
Plecoptera	*Chloroperla* sp.
Plecoptera	*Xanthoperla apicalis*
Plecoptera	*Capnia* sp.
Plecoptera	*Leuctra* sp.
Plecoptera	*Nemoura* sp.
Plecoptera	*Nemurella pictetii*
Plecoptera	*Protonemura* sp.
Plecoptera	*Dinocras cephalotes*
Plecoptera	*Perla* sp.
Plecoptera	*Marthamea vitripennis*
Plecoptera	*Isoperla* sp.
Plecoptera	*Perlodes* sp.
Plecoptera	*Brachyptera* sp.
Plecoptera	*Taeniopteryx nebulosa*
Odonata	*Calopteryx splendens*
Odonata	*Cercion lindenii*
Odonata	*Enallagma cyathigerum*
Odonata	*Ischnura elegans*
Odonata	Coenagrionidae Gen. sp.
Odonata	*Cordulegaster bidentata*
Odonata	*Cordulegaster heros*
Odonata	*Gomphus* sp.
Odonata	*Gomphus vulgatissimus*
Odonata	*Gomphus flavipes*
Odonata	*Onychogomphus forcipatus*
Odonata	*Ophiogomphus cecilia*
Odonata	*Orthetrum brunneum*
Odonata	*Platycnemis pennipes*
Heteroptera	*Aphelocheirus aestivalis*
Heteroptera	Corixinae Gen. sp.
Heteroptera	*Micronecta* sp.
Megaloptera	*Sialis fuliginosa*

Table A2. *Cont.*

Higher Taxon	Taxon
Megaloptera	*Sialis lutaria*
Megaloptera	*Sialis nigripes*
Hymenoptera	*Agriotypus armatus*
Coleoptera	*Bidessus* sp. Ad.
Coleoptera	*Platambus maculatus* Ad.
Coleoptera	*Elmis* sp. Ad.
Coleoptera	*Elmis* sp. Lv.
Coleoptera	*Esolus* sp. Ad.
Coleoptera	*Esolus* sp. Lv.
Coleoptera	*Limnius* sp. Ad.
Coleoptera	*Limnius* sp. Lv.
Coleoptera	*Normandia nitens* Ad.
Coleoptera	*Oulimnius* sp. Ad.
Coleoptera	*Oulimnius* sp. Lv.
Coleoptera	*Riolus* sp. Ad.
Coleoptera	*Riolus* sp. Lv.
Coleoptera	*Stenelmis canaliculata* Ad.
Coleoptera	*Orectochilus villosus* Lv.
Coleoptera	*Haliplus* sp. Ad.
Coleoptera	*Haliplus* sp. Lv.
Coleoptera	*Helophorus* sp. Ad.
Coleoptera	*Hydraena* sp. Ad.
Coleoptera	*Ochthebius* sp. Ad.
Trichoptera	*Brachycentrus montanus*
Trichoptera	*Brachycentrus subnubilus*
Trichoptera	*Ecnomus tenellus*
Trichoptera	*Agapetus* sp.
Trichoptera	*Agapetus laniger*
Trichoptera	*Glossosoma boltoni*
Trichoptera	*Glossosoma conformis*
Trichoptera	*Glossosoma intermedium*
Trichoptera	*Goera pilosa*
Trichoptera	*Silo nigricornis*
Trichoptera	*Silo piceus*
Trichoptera	*Cheumatopsyche lepida*
Trichoptera	*Hydropsyche bulbifera*
Trichoptera	*Hydropsyche bulgaromanorum*
Trichoptera	*Hydropsyche contubernalis*
Trichoptera	*Hydropsyche incognita*
Trichoptera	*Hydropsyche modesta*
Trichoptera	*Hydropsyche ornatula*
Trichoptera	*Hydropsyche pellucidula*
Trichoptera	*Hydropsyche siltalai*
Trichoptera	*Hydropsyche* sp.
Trichoptera	*Hydroptila* sp.
Trichoptera	*Orthotrichia* sp.
Trichoptera	*Lepidostoma hirtum*
Trichoptera	*Athripsodes albifrons*
Trichoptera	*Athripsodes cinereus*
Trichoptera	*Athripsodes* sp.
Trichoptera	*Ceraclea annulicornis*
Trichoptera	*Ceraclea dissimilis*
Trichoptera	*Mystacides azurea*
Trichoptera	*Mystacides longicornis*
Trichoptera	*Mystacides nigra*
Trichoptera	*Oecetis lacustris*
Trichoptera	*Oecetis notata*

Table A2. *Cont.*

Higher Taxon	Taxon
Trichoptera	*Setodes punctatus*
Trichoptera	*Setodes* sp.
Trichoptera	*Anabolia furcata*
Trichoptera	*Chaetopteryx* sp.
Trichoptera	*Halesus digitatus*
Trichoptera	*Halesus radiatus*
Trichoptera	Limnephilinae Gen. sp.
Trichoptera	*Limnephilus extricatus*
Trichoptera	*Potamophylax rotundipennis*
Trichoptera	*Potamophylax* sp.
Trichoptera	*Philopotamus* sp.
Trichoptera	*Cyrnus trimaculatus*
Trichoptera	*Polycentropus flavomaculatus*
Trichoptera	*Lype reducta*
Trichoptera	*Psychomyia pusilla*
Trichoptera	*Tinodes* sp.
Trichoptera	*Rhyacophila* s. str. sp.
Trichoptera	*Notidobia ciliaris*
Trichoptera	*Sericostoma* sp.
Diptera	*Limnophora* sp.
Diptera	*Lispe* sp.
Diptera	*Atherix ibis*
Diptera	*Ibisia marginata*
Diptera	*Ibisia* sp.
Diptera	*Liponeura* sp.
Diptera	Ceratopogoninae Gen. sp.
Diptera	*Dasyhelea* sp.
Diptera	*Brillia bifida*
Diptera	Chironomini Gen. sp.
Diptera	*Chironomus obtusidens*-Gr.
Diptera	*Chironomus plumosus*-Gr.
Diptera	*Chironomus thummi*-Gr.
Diptera	*Chironomus plumosus*
Diptera	*Chironomus* sp.
Diptera	*Corynoneura* sp.
Diptera	Orthocladiinae Gen. sp.
Diptera	Diamesinae Gen. sp.
Diptera	*Monodiamesa* sp.
Diptera	Orthocladiinae Gen. sp.
Diptera	*Paratendipes* sp.
Diptera	*Potthastia longimana*-Gr.
Diptera	*Procladius* sp.
Diptera	*Prodiamesa olivacea*
Diptera	*Prodiamesa rufovittata*
Diptera	Tanypodinae Gen. sp.
Diptera	Tanytarsini Gen. sp.
Diptera	*Thienemanniella* sp.
Diptera	Dolichopodidae Gen. sp.
Diptera	Clinocerinae Gen. sp.
Diptera	Hemerodromiinae Gen. sp.
Diptera	*Antocha* sp.
Diptera	Chioneinae Gen. sp.
Diptera	*Hexatoma* sp.
Diptera	Limnophilinae Gen. sp.
Diptera	Limoniinae Gen. sp.
Diptera	*Dicranota* sp.

Table A2. *Cont.*

Higher Taxon	Taxon
Diptera	*Pedicia* sp.
Diptera	Psychodidae Gen. sp.
Diptera	Psychodidae Gen. sp.
Diptera	Psychodidae Gen. sp.
Diptera	*Ptychoptera* sp.
Diptera	*Prosimulium* sp.
Diptera	*Simulium* sp.
Diptera	Syrphidae Gen. sp.
Diptera	*Chrysops* sp.
Diptera	*Tabanus* sp.
Diptera	*Tipula* sp.
Lepidoptera	*Nymphula stagnata*

Appendix C

Table A3. Statistically significant Spearman's correlation coefficients (RSp) for the combinations of typology variables with variables from all groups (** P < 0.01, * P < 0.05); |RSp| > 0.50 are in bold. See Table 2 for environmental variable codes. Med–median.

	Depth_Mean	C_Size	Slope	Altitude	Substrat_Code
Depth_mean		0.494**	**−0.613****		−0.373**
C_size	0.494**		−0.405**		−0.465**
slope	**−0.613****	−0.405**		0.299**	0.337**
altitude			0.299**		0.499**
substrat_code	−0.373**	−0.465**	0.337**	0.499**	
T_med	0.197*			−0.278**	−0.361**
pH_med		−0.277**			
cond_med	−0.461**	−0.418**		−0.631**	
DO_med	−0.257**			−0.209*	
DOsat_med					
TSS_med	0.329**	0.391**	−0.233*		−0.293**
KPK_Cmed	−0.313**			−0.278**	−0.277**
BPK5_med	−0.221*			−0.443**	−0.248*
PO4_P_med	**−0.505****	**−0.561****	0.243*	−0.353**	
Ntot_med	**−0.643****	−0.427**	0.401**		
NH4_N_med	−0.213*	−0.404**			
NO2_N_med	**−0.613****	**−0.573****	0.461**		0.439**
NO3_N_med	**−0.607****	−0.444**	0.414**		
Q	**0.591****	**0.564****	**−0.509****	−0.198*	−0.450**
Qnp	**0.571****	**0.520****	**−0.513****	−0.250*	−0.488**
Qs	**0.544****	0.443**	**−0.546****	−0.378**	−0.490**
Qvk	0.336**		−0.382**	−0.487**	−0.383**
RHQ	**−0.652****		0.350**	−0.330**	
RHM				0.299**	0.254**
HLM	−0.395**			**−0.685****	−0.437**
HMM	−0.392**			**−0.582****	−0.410**
HQM	−0.408**			**−0.625****	−0.408**
C_urb	**−0.524****	−0.366**	0.243*		
C_nat	0.205*			**0.922****	0.491**
C_agrE	−0.327**	−0.444**		**−0.703****	
C_agrI1		0.467**			
C_argI2			−0.214*	**−0.879****	−0.398**

Appendix D

Table A4. Statistically significant Spearman's correlation coefficients (RSp) for the combinations of water quality variables with variables from all groups (** P < 0.01, * P < 0.05); |RSp| > 0.50 are in bold. See Table 2 for environmental variable codes. Med–median.

	T_med	pHMed	Cond_Med	DO_Med	DOsat_Med	TSS_Med	KPK_Cmed	BPK5_Med	PO4_P_Med	Ntot_Med	NH4_N_Med	NO2_N_Med	NO3_N_Med
Depth_mean	0.197*		−0.461**	−0.257**		0.329**	−0.313**	−0.221*	**−0.505**	**−0.643**	−0.213*	**−0.613**	**−0.607**
C_size		−0.277**	−0.418**			0.391**			**−0.561**	−0.427**	−0.404**	**−0.573**	−0.444**
slope	−0.278**					−0.233*			0.243*	0.401**		0.461**	0.414**
altitude	−0.361**		**−0.631**	−0.209*			−0.278**	−0.443**	−0.353**				
substrat_code						−0.293**	−0.277**	−0.248*				0.439**	
T_med			0.314**										
pH_med			0.261**										−0.205*
cond_med	0.314**	0.261**				−0.306**		0.381**	**0.654**	0.418**		0.222*	0.410**
DO_med					**0.570**								
DOsat_med				**0.570**						−0.273**	−0.381**	−0.220*	−0.211*
TSS_med			−0.306**						−0.324**			−0.259**	
KPK_Cmed								**0.597**	0.420**	0.484**	0.278**	0.205*	0.382**
BPK5_med			0.381**				**0.597**		0.481**	0.323**	0.194*		0.259**
PO4_P_med			**0.654**			−0.324**	0.420**	0.481**		**0.664**	0.496**	**0.667**	**0.687**
Ntot_med			0.418**		−0.273**		0.484**	0.323**	**0.664**		0.423**	**0.754**	**0.945**
NH4_N_med					−0.381**		0.278**	0.194*	0.496**	0.423**		**0.502**	0.377**
NO2_N_med			0.222*		−0.220*	−0.259**	0.205*		**0.667**	**0.754**	**0.502**		**0.817**
NO3_N_med		−0.205*	0.410**		−0.211*		0.382**	0.259**	**0.687**	**0.945**	0.377**	**0.817**	
Q			−0.293**			0.340**			−0.321**	−0.417**	−0.250*	**−0.529**	−0.397**
Qnp			−0.253**			0.360**			−0.312**	−0.434**	−0.212*	**−0.554**	−0.456**
Qs						0.287**			−0.214*	−0.439**	−0.283**	**−0.555**	−0.436**
Qvk	0.336**	0.250*	0.281**	0.307**					−0.251*	−0.265**	−0.379**	−0.237*	
RHQ	−0.214*		0.306**	0.207*			0.358**	0.231*	0.249*	0.395**		0.245*	0.332**
RHM			−0.216*									0.224*	
HLM			**0.521**	0.248*			0.480**	0.407**	0.313**	0.284**			0.224*
HMM			**0.552**	0.250*			0.275**	0.338**	0.280**				
HQM			**0.542**	0.249*			0.332**	0.390**	0.303**	0.210*			
C_urb					−0.224*		**0.573**	0.335**	**0.500**	**0.689**	0.442**	**0.626**	**0.637**
C_nat	−0.314**	−0.199*	**−0.752**	−0.203*			−0.356**	−0.467**	**−0.508**	−0.202*			
C_agrE	0.317**	0.253**	**0.838**			−0.299**	0.235*	0.387**	**0.630**	0.279**	0.305**		0.233*
C_agrI1	−0.340**	−0.252**	**−0.519**		−0.232*	0.321**	**0.503**						
C_argI2	0.415**		**0.785**		0.212*	−0.211*		0.384**	**0.526**				

Appendix E

Table A5. Statistically significant Spearman's correlation coefficients (RSp) for the combinations of land use variables with variables from all groups (** P < 0.01, * P < 0.05); |RSp| > 0.50 are in bold. See Table 2 for environmental variable codes. Med–median.

	C_urb	C_Nat	C_AgrI	C_AgrE	C_AgrI2
Depth_mean	−0.524**	0.205*		−0.327**	
C_size	−0.366**			−0.444**	
slope	0.243*		−0.286**		−0.214*
altitude		0.922**	−0.948**	−0.703**	−0.879**
substrat_code		0.491**	−0.507**		−0.398**
T_med		−0.314**	0.323**	0.317**	0.415**
pH_med		−0.199*		0.253**	
cond_med		−0.752**	0.629**	0.838**	0.785**
DO_med		−0.203*			
DOsat_med	−0.224*				0.212*
TSS_med				−0.299**	−0.211*
KPK_Cmed	0.573**	−0.356**	0.328**	0.235*	
BPK5_med	0.335**	−0.467**	0.440**	0.387**	0.384**
PO4_P_med	0.500**	−0.508**	0.423**	0.630**	0.526**
Ntot_med	0.689**	−0.202*		0.279**	
NH4_N_med	0.442**			0.305**	
NO2_N_med	0.626**				
NO3_N_med	0.637**			0.233*	
Q	−0.244*		0.223*		
Qnp	−0.328**		0.286**		
Qs	−0.372**	−0.312**	0.423**		0.337**
Qvk	−0.383**	−0.483**	0.530**	0.378**	0.565**
RHQ	0.239*	−0.272**	0.251*	0.218*	
RHM	0.208*	0.208*	−0.221*		−0.238*
HLM	0.204*	−0.697**	0.591**	0.499**	0.483**
HMM		−0.604**	0.491**	0.464**	0.464**
HQM		−0.638**	0.545**	0.473**	0.479**
C_urb					
C_nat			−0.948**	−0.841**	−0.917**
C_agrE		−0.841**	0.740**		0.858**
C_agrI1	0.492**			−0.482**	−0.391**
C_argI2		−0.917**	0.928**	0.858**	

Appendix F

Table A6. Statistically significant Spearman's correlation coefficients (RSp) for the combinations of hydromorphology variables with variables from all groups (** P < 0.01, * P < 0.05); |RSp| > 0.50 are in bold. See Table 2 for environmental variable codes. Med–median.

	Q	Qnp	Qs	Qvk	RHQ	RHM	HLM	HMM	HQM
Depth_mean	**0.591****	**0.571****	**0.544****	0.336**	**−0.652****		−0.395**	−0.392**	−0.408**
C_size	**0.564****	**0.520****	0.443**						
slope	**−0.509****	**−0.513****	**−0.546****	−0.382**	0.350**				
altitude	−0.198*	−0.250*	−0.378**	−0.487**	−0.330**	0.299**	**−0.685****	**−0.582****	**−0.625****
substrat_code	−0.450**	−0.488**	−0.490**	−0.383**		0.254**	−0.437**	−0.410**	−0.408**
T_med				0.336**	−0.214*				
pH_med				0.250*					
cond_med	−0.293**	−0.253**		0.281**	0.306**	−0.216*	**0.521****	**0.552****	**0.542****
DO_med					0.207*		0.248*	0.250*	0.249*
DOsat_med				0.307**					
TSS_med	0.340**	0.360**	0.287**						
KPK_Cmed					0.358**		0.480**	0.275**	0.332**
BPK5_med					0.231*		0.407**	0.338**	0.390**
PO4_P_med	−0.321**	−0.312**	−0.214*		0.249*		0.313**	0.280**	0.303**
Ntot_med	−0.417**	−0.434**	−0.439**	−0.251*	0.395**		0.284**		0.210*
NH4_N_med	−0.250*	−0.212*	−0.283**	−0.265**					
NO2_N_med	**−0.529****	**−0.554****	**−0.555****	−0.379**	0.245*	0.224*			
NO3_N_med	−0.397**	−0.456**	−0.436**	−0.237*	0.332**		0.224*		
Q		**0.806****	**0.780****	0.476**	−0.336**				
Qnp	**0.806****		**0.945****	**0.649****	−0.273**				
Qs	**0.780****	**0.945****		**0.800****	−0.317**				
Qvk	0.476**	**0.649****	**0.800****		−0.247*				
RHQ	−0.336**	−0.273**	−0.317**	−0.247*		−0.338**	**0.560****	**0.511****	**0.610****
RHM					−0.338**		−0.355**	**−0.667****	**−0.629****
HLM					**0.560****	−0.355**		**0.842****	**0.887****
HMM					**0.511****	**−0.667****	**0.842****		**0.965****
HQM					**0.610****	**−0.629****	**0.887****	**0.965****	
C_urb	−0.244*	−0.328**	−0.372**	−0.383**	0.239*	0.208*	0.204*		
C_nat			−0.312**	−0.483**	−0.272**	0.208*	**−0.697****	**−0.604****	**−0.638****
C_agrE				0.378**	0.218*		0.499**	0.464**	0.473**
C_agrI1	0.244*			−0.391**	0.263**				
C_argI2			0.337**	**0.565****		−0.238*	0.483**	0.464**	0.479**

References

1. EU. *Directive 2000/60/EC of the European Parliament and of the Council Establishing a Framework for the Community Action in the Field of Water Policy*; EU: Brussels, Belgium, 2000; p. 72. Available online: https://eur-lex.europa.eu/legal-content/EN/TXT/?uri=CELEX:32000L0060 (accessed on 25 November 2019).
2. Sweeting, R.A. River Pollution. In *The Rivers Handbook*; Hydrological and Ecological Principles; Calow, P., Petts, G.E., Eds.; Blackwell Scientific: Oxford, UK, 1994; pp. 23–32.
3. Angradi, T.R.; Pearson, M.S.; Jicha, T.M.; Taylor, D.L.; Bolgrien, D.W.; Moffett, M.F.; Blocksom, K.A.; Hill, B.H. Using stressor gradients to determine reference expectations for great river fish assemblages. *Ecol. Indic.* **2009**, *9*, 748–764. [CrossRef]
4. Schinegger, R.; Trautwein, C.; Melcher, A.; Schmutz, S. Multiple human pressures and their spatial patterns in European running waters. *Water Environ. J.* **2012**, *26*, 261–273. [CrossRef] [PubMed]
5. Jackson, M.C.; Loewen, C.J.G.; Vinebrooke, R.D.; Chimimba, C.T. Net effects of multiple stressors in freshwater ecosystems: A meta-analysis. *Glob. Chang. Biol.* **2016**, *22*, 180–189. [CrossRef] [PubMed]
6. Nõges, P.; Argillier, C.; Borja, Á.; Garmendia, J.M.; Hanganu, J.; Kodeš, V.; Pletterbauer, F.; Saguis, A.; Birk, S. Quantified biotic and abiotic responses to multiple stress in freshwater, marine and ground waters. *Sci. Total Environ.* **2016**, *540*, 43–52. [CrossRef]
7. Vitousek, P.M.; Mooney, H.A.; Lubchenco, J.; Melillo, J.M. Human Domination of Earth's Ecosystems. *Science* **1997**, *277*, 494–499. [CrossRef]
8. Allan, J.D.; Castillo, M.M. *Stream Ecology: Structure and Function of Running Waters*, 2nd ed.; Springer: Amsterdam, The Netherlands, 2007; p. 436.
9. Petts, G.E.; Möller, H.; Roux, A.L. *Historical Change of Large Alluvial Rivers: Western Europe*; John Wiley & Sons: Chichester, UK, 1989; p. 355.
10. Aarts, B.G.W.; Van den Brink, F.W.B.; Nienhuis, P.H. Habitat loss as the main cause of the slow recovery of fish faunas of regulated large rivers in Europe: The transversal floodplain gradient. *River Res. Appl.* **2004**, *20*, 3–23. [CrossRef]
11. Petts, G.E. *Impounded Rivers: Perspectives for Ecological Management*; John Wiley & Sons: Chichester, UK, 1984; p. 326.
12. Nilsson, C.; Reidy, C.A.; Dynesius, M.; Revenga, C. Fragmentation and flow regulation of the world's large river systems. *Science* **2005**, *308*, 405–408. [CrossRef]
13. Zarfl, C.; Lumsdonm, A.E.; Berlekampm, J.; Tydecks, L.; Tockner, K.; Zarfl. A global boom in hydropower dam construction. *Aquat. Sci.* **2014**, *77*, 161–170. [CrossRef]
14. Ward, J.V.; Tockner, K.; Uehlinger, U.; Malard, F. Understanding natural patterns and processes in river corridors as the basis for effective river restoration. *Regul. River* **2001**, *17*, 311–323. [CrossRef]
15. Davies, S.P.; Jackson, S.K. The biological condition gradient: A descriptive model for interpreting change in aquatic ecosystems. *Ecol. Appl.* **2006**, *16*, 1251–1266. [CrossRef]
16. Tockner, K.; Uehlinger, U.; Robinson, C.T.; Tonolla, D.; Siber, R.; Peter, F.D. Introduction to European Rivers. In *Rivers of Europe*, 1st ed.; Tockner, K., Robinson, C.T., Uehlinger, U., Eds.; Academic Press: London, UK, 2008; pp. 1–21.
17. Urbanič, G. Hydromorphological degradation impact on benthic invertebrates in large rivers in Slovenia. *Hydrobiologia* **2014**, *729*, 191–207. [CrossRef]
18. Schwarz, U. Pilot Study: Hydromorphological survey and mapping of the Drava and Mura Rivers. IAD-Report Prepared by FLUVIUS, Floodplain Ecology and River Basin Management. 2007. Available online: https://www.danube-iad.eu/docs/reports/HydromorphIAD_Mura_Drava2007.pdf (accessed on 25 November 2019).
19. Schneider-Jacoby, M. The Sava and Drava floodplains: Threatened ecosystems of international importance. *Arch. Hydrobiol. Large Rivers* **2005**, *16* (Suppl. 158), 249–288. [CrossRef]
20. CLC. *CORINE Land Cover 2012*; European Environment Agency: Copenhagen, Denmark, 2012.
21. Tavzes, B.; Urbanič, G. New indices for assessment of hydromorphological alteration of rivers and their evaluation with benthic invertebrate communities; Alpine case study. *Rev. Hydrobiol.* **2009**, *2*, 133–161.
22. Petkovska, V.; Urbanič, G.; Mikoš, M. Variety of the guiding image of rivers-defined for ecologically relevant habitat features at the meeting of the Alpine, Mediterranean, lowland and karst regions. *Ecol. Eng.* **2015**, *81*, 373–386. [CrossRef]

23. Raven, P.J.; Holmes, N.T.H.; Dawson, F.H.; Fox, P.J.A.; Everard, M.; Fozzard, I.R.; Rouen, K.J. River Habitat Survey in Britain and Ireland Field Survey Guidance manual: Version 2003. 2003. Available online: https://assets.publishing.service.gov.uk/government/uploads/system/uploads/attachment_data/file/311579/LIT_1758.pdf (accessed on 22 February 2020).

24. Raven, P.J.; Holmes, N.T.H.; Dawson, F.H.; Fox, P.J.A.; Everard, M.; Fozzard, I.R.; Rouen, K.J. *River Habitat Quality the Physical Character of Rivers and Streams in the UK and Isle of Man*; (River Habitat survey Report No. 2.); Environment Agency: Bristol, UK, 1998; p. 86.

25. Pavlin, M.; Birk, S.; Hering, D.; Urbanič, G. The role of land use, nutrients, and other stressors in shaping benthic invertebrate assemblages in Slovenian rivers. *Hydrobiologia* **2011**, *678*, 137–153. [CrossRef]

26. Urbanič, G. Ecological status assessment of rivers in Slovenia—An overview. *Nat. Slov.* **2011**, *13*, 5–16.

27. AQEM Consortium. Manual for the application of the AQEM system. A Comprehensive Method to Assess European Streams Using Benthic Macroinvertebrates, Developed for the Purpose of the Water Framework Directive. Version 1. 2002, p. 198. Available online: http://www.aqem.de/mains/products.php (accessed on 22 February 2020).

28. Urbanič, G.; Toman, M.J.; Krušnik, C. Microhabitat type selection of caddisfly larvae (Insecta: Trichoptera) in a shallow lowland stream. *Hydrobiologia* **2005**, *541*, 1–12. [CrossRef]

29. Petkovska, V.; Urbanič, G. Effect of fixed-fraction subsampling on macroinvertebrate bioassessment of rivers. *Environ. Monitor. Assess.* **2010**, *169*, 179–201. [CrossRef]

30. Ter Braak, C.J.F.; Šmilauer, P. *Canoco Reference Manual and User's Guide: Software for Ordination, Version 5.0*; Microcomputer Power: Ithaca, NY, USA, 2012; p. 496.

31. Lepš, J.; Šmilauer, P. *Multivariate Analysis of Ecological Data using CANOCO*, 2nd ed.; Cambridge University Press: Cambridge, UK, 2014; p. 362.

32. IBM. IBM SPSS Statistics 21 Core System User's Guide; IBM Corporation: 2012. Available online: http://www.sussex.ac.uk/its/pdfs/SPSS_Core_System_Users_Guide_21.pdf (accessed on 22 February 2020).

33. Hill, M.O.; Gauch, H.G., Jr. Detrended correspondence analysis: An improved ordination technique. *Vegetatio* **1980**, *42*, 47–58. [CrossRef]

34. Ter Braak, C.J.F. Canonical correspondence analysis: A new eigenvector technique for multivariate direct gradient analysis. *Ecology* **1986**, *67*, 1176–1179. [CrossRef]

35. Borcard, D.; Legendre, P.; Drapeau, P. Partialling out the spatial component of ecological variation. *Ecology* **1992**, *73*, 1045–1055. [CrossRef]

36. Allan, J.D. Landscapes and riverscapes: The influence of land use on stream ecosystems. *Ann. Rev. Ecol. Evol. Syst.* **2004**, *35*, 257–284. [CrossRef]

37. Bonada, N.; Prat, N.; Resh, V.H.; Statzner, B. Developments in aquatic insect biomonitoring: A comparative analysis of recent approaches. *Ann. Rev. Entomol.* **2006**, *51*, 495–523. [CrossRef] [PubMed]

38. Urbanič, G.; Toman, M.J. Influence of environmental variables on stream caddis larvae in three Slovenian ecoregions: Alps, Dinaric Western Balkans and Pannonian lowland. *Inter. Rev. Hydrobiol.* **2007**, *92*, 582–602. [CrossRef]

39. Smolar-Žvanut, N.; Mikoš, M. The impact of flow regulation by hydropower dams on the periphyton community in the Soča River, Slovenia. *Hydrol. Sci. J.* **2014**, *59*, 1032–1045. [CrossRef]

40. Downes, B.J. Back to the future: Little-used tools and principles of scientific inference can help disentangle effects of multiple stressors on freshwater ecosystems. *Freshw. Biol.* **2010**, *55*, 60–79. [CrossRef]

41. Wetzel, R. *Limnology: Lake and River Ecosystems*, 3rd ed.; Academic Press: San Diego, CA, USA, 2001; p. 1006.

42. Donohue, I.; McGarrigle, M.L.; Mills, P. Linking catchment characteristics and water chemistry with the ecological status of Irish rivers. *Wat. Res.* **2006**, *40*, 91–98. [CrossRef]

43. Petkovska, V.; Urbanič, G. The links between morphological parameters and benthic invertebrate assemblages, and general implications for hydromorphological river management. *Ecohydrology* **2015**, *8*, 67–82. [CrossRef]

44. Urbanič, G.; Petkovska, V.; Šiling, R.; Knehtl, M. Knowing impacts and accepting solutions can lead to better water ecological status. In Proceedings of the Water days 2018, Portorož, Slovenia, 18–19, October 2018; Cerkvenik, S., Rojnik, E., Eds.; Slovensko društvo za zaščito voda: Ljubljana, Slovenia, 2018; pp. 45–54.

45. Meštrov, M.; Dešković, I.; Tavčar, V. Saprobiological and physico-chemical researches of the Sava River in the course of several years. *Bull. Scient. Cons. Academ. Sci. Arts Yougoslav.* **1976**, *21*, 144–145.

46. Mihaljević, Z.; Kerovec, M.; Tavčar, V.; Bukvić, I. Macroinvertebrate community on an artificial substrate in the Sava river: Long-term changes in the community structure and water quality. *Biologia* **1998**, *53*, 611–620.

47. EEA. *European Waters—Assessment of Status and Pressures 2018, EEA Report No 7/2018*; European Environment Agency: Copenhagen, Denmark, 2018; p. 85.

48. Ferrier, S.; Guisan, A. Spatial modelling of biodiversity at the community level. *J. Appl. Ecol.* **2006**, *43*, 393–404. [CrossRef]

49. Olden, J.; Joy, M.; Death, R. Rediscovering the species in community-wide predictive modelling. *Ecol. Appl.* **2006**, *16*, 1449–1460. [CrossRef]

50. Friberg, N.; Skriver, J.; Larsen, S.E.; Pedersen, M.L.; Buffagni, A. Stream macroinvertebrate occurrence along gradients in organic pollution and eutrophication. *Freshw. Biol.* **2009**, *55*, 1405–1419. [CrossRef]

51. Ormerod, S.J.; Dobson, M.; Hildrew, A.G.; Townsend, C.R. Multiple stressors in freshwater ecosystems. *Freshw. Biol.* **2010**, *55*, 1–4. [CrossRef]

52. Guisan, A.; Lehmann, A.; Ferrier, S.; Austin, M.; Overton, J.M.C.; Aspinall, R.; Hastie, T. Making better biogeographical predictions of species' distributions. *J. Appl. Ecol.* **2006**, *43*, 386–392. [CrossRef]

53. Sladeček, V. System of Water Quality. *Arch. Hydrobiol.Ergeb. Limnol.* **1973**, *7*, 1–218. [CrossRef]

54. Cairns, J.; Pratt, J.R. A history of biological monitoring using benthic macroinvertebrates. In *Freshwater Biomonitoring and Benthic Macroinvertebrates*, 1st ed.; Rosenberg, D.M., Resh, V.H., Eds.; Chapman and Hall: New York, NY, USA, 1993; pp. 10–27.

55. Zelinka, M.; Marvan, P. Zur Präzisierung der biologischen Klassifikation der Reinheit fließender Gewässer. *Arch. Hydrobiol.* **1961**, *57*, 389–407.

56. Johnson, B.L.; Richardson, W.B.; Naimo, T.J. Past, present, and future concepts in large river ecology. *BioScience* **1995**, *45*, 134–141. [CrossRef]

57. Tavzes, B.; Urbanič, G.; Toman, M.J. Biological and hydromorphological integrity of the small urban stream. *Phys. Chem. Earth. (2002)* **2006**, *31*, 1062–1074. [CrossRef]

58. Friberg, N.; Bonada, N.; Bradley, D.C.; Dunbar, M.J.; Edwards, F.K.; Grey, J.; Hayes, R.B.; Hildrew, A.G.; Lamouroux, N.; Trimmer, M.; et al. Biomonitoring of human impacts in freshwater ecosystems: The good, the bad and the ugly. *Adv. Ecol. Res.* **2011**, *44*, 1–68.

59. Petkovska, V.; Urbanič, G. The links between river morphological variables and benthic invertebrate assemblages: Comparison among three European ecoregions. *Aquat. Ecol.* **2015**, *49*, 59–173. [CrossRef]

60. Knehtl, M.; Petkovska, V.; Urbanič, G. Is it time to eliminate field surveys from hydromorphological assessments of rivers?—Comparison between a field survey and remote sensing approach. *Ecohydrology* **2018**, *11*, 1–12. [CrossRef]

Article

Inconsistent Relationships of Primary Consumer N Stable Isotope Values to Gradients of Sheep/Beef Farming Intensity and Flow Reduction in Streams

Katharina Lange *, Colin R. Townsend and Christoph D. Matthaei

Department of Zoology, University of Otago, Dunedin 9016, New Zealand; colin.townsend@otago.ac.nz (C.R.T.); christoph.matthaei@otago.ac.nz (C.D.M.)

* Correspondence: katha.lange@gmail.com; Tel.: +41-78-824-0584

Received: 14 August 2019; Accepted: 22 October 2019; Published: 26 October 2019

Abstract: Stable isotope values of primary consumers have been proposed as indicators of human impacts on nitrogen dynamics. Until now, these values have been related only to single-stressor gradients of land-use intensity in stream ecology, whereas potential interactive effects of multiple stressors are unknown. It also remains unknown whether stable isotope values of different primary consumers show similar relationships along gradients of stressor intensities. We sampled three common invertebrate grazers along gradients of sheep/beef farming intensity (0–95% intensively managed exotic pasture) and flow reduction (0–92% streamflow abstracted for irrigation). The $\delta^{15}N$ values of the three primary consumers differed substantially along stressor gradients. *Deleatidium* $\delta^{15}N$ values were positively related to farming intensity, showing a saturation curve, whereas *Physella* snail $\delta^{15}N$ values were negatively related to farming intensity and *Potamopyrgus* snail $\delta^{15}N$ values showed no relationship. In addition, *Deleatidium* stable isotope values responded positively to flow reduction intensity, a previously unstudied variable. An antagonistic multiple-stressor interaction was detected only for the mayfly *Deleatidium*, which occurred in streams experiencing up to 53% farming intensity. The lack of consistency in the relationships of the most important primary consumer grazers along the studied gradients may reduce their suitability as an indicator of anthropogenic N inputs.

Keywords: agricultural land use; antagonism; *Deleatidium*; grazer-scrapers

---

1. Introduction

Most ecosystems are exposed to multiple environmental stressors acting simultaneously [1,2] and combined multiple-stressor effects can be difficult to predict because they can be larger or smaller than expected based on the effects of the individual stressors involved [3]. Stable isotope values have long been considered indicators of anthropogenic disturbance [4,5]. In aquatic environments, the use of $\delta^{15}N$ (a measure of the abundance of the heavier isotope ^{15}N relative to the lighter isotope ^{14}N) as potential indicators of land-use intensification and nitrogen enrichment has been suggested for streams and rivers [6–8], ponds and lakes [9,10], estuaries [11] and coastal waters [12,13]. Agricultural land use can impose a variety of stressors on stream ecosystems, including nutrient enrichment, increased sediment load, higher light availability and augmented water temperatures due to removal of riparian vegetation [14,15]. Another important stressor is water diversion for irrigation, which commonly reduces stream discharge and flow velocity and changes sediment and temperature regimes [16]. Any of these stressors may alter stream nitrogen dynamics, for example by flow reduction intensifying in-stream nitrogen transformation processes such as denitrification by increasing water retention times, respiration rates and occurrence of anoxic conditions [17]. However, the combined effects of multiple agricultural stressors on stable isotope values in stream ecosystems still need to be investigated.

Delta ^{15}N values differ between nitrogen sourced from precipitation, fertilizer and sewage [18,19], and also due to subsequent fractionation during assimilation by plants and other nitrogen transformation processes [20–22]. Fractionation is the enrichment of one isotope relative to another through biogeochemical processes that discriminate against the heavier isotope. Further, gaseous nitrogen losses through denitrification and ammonia volatilisation can lead to even more elevated δ^{15}N values of the nitrogen initially derived from both inorganic fertilizer and manure [23]. Catchment land-use intensity has been positively related to δ^{15}N values of stream water [19,24,25], sediment [26], aquatic plants [27], consumers [6,7,28] and fish [6].

Among stream consumers, invertebrate grazers represent an important link for nutrient and energy transfer from primary producers to higher trophic levels. Moreover, they have been used as bioindicators because they are widely distributed in running waters and generally reflect the isotopic values of their periphyton food sources (after taking into account enrichment through trophic fractionation) [29]. Compared to the high spatiotemporal variation of periphyton stable isotope values [30], grazers provide a more time-integrated and habitat-integrated measure of nitrogen sources and transformation processes [5,31,32].

Most of the existing stream studies have reported positive linear relationships between δ^{15}N values and catchment land-use intensity [18,28], although Larson et al. [8] reported a saturation curve for δ^{15}N values along a wide gradient of % agriculture. To our knowledge, no studies have investigated the combined effects of multiple, simultaneously operating agricultural stressors on consumer stable isotope ratios, even though non-additive effects such as antagonisms and synergisms have been documented from experiments assessing changes in diversity and ecosystem processes in freshwater ecosystems [33]. Nor have previous studies focused on multiple primary consumers: most have investigated the relationship between a single primary consumer taxon or calculated δ^{15}N values from several taxa combined where the focal consumer taxon was not present at all sites [7,8]. Whether different primary consumer taxa show similar relationships along gradients of anthropogenic disturbance remains to be investigated.

Against this background, we studied consumer stable isotope values at 43 stream sites spanning wide gradients of land-use intensity (sheep/beef farming intensity and streamflow reduction) within an agricultural river catchment, to test the following hypotheses:

1. δ^{15}N values of common primary consumers (grazers) will show similar patterns along the gradients of catchment land-use intensity because they all feed on one resource (periphyton) and are therefore ingesting food with the same isotopic composition;
2. δ^{15}N values are positively related to catchment land-use intensities because higher farming intensity and greater flow reduction increase the input of heavy isotopes from fertilisation and increase the intensity of nitrogen transformation processes [17]; and
3. δ^{15}N values follow antagonistic response patterns along the gradients of land-use intensity and flow reduction because both stressors have strong individual effects and their joint effects cannot exceed 100% [3].

2. Methods

2.1. Study Sites

The 43 sites were selected within a single catchment, the Manuherikia River in Central Otago, which is among the driest in New Zealand. All sites experienced similar climatic conditions. The sites included 3rd–5th order streams (subcatchment sizes ranged from 2.7 to 367.9 km^2) chosen to provide wide gradients of both % Farming Intensity and % Flow Reduction (Figure 1, further site details in [34]). The catchment contains sites with high flow reduction intensity in areas of low sheep/beef farming intensity and vice versa because water for irrigation is transported in water races from one sub-catchment to another. Our strategic selection of sites along the two land-use gradients ensured that % Farming Intensity and % Flow Reduction were uncorrelated (R^2 = 0.03, Figure 2).

Figure 1. Locations of the 43 study sites in the Manuherikia River catchment.

Land cover in the Manuherikia River catchment is dominated by sheep and beef-cattle pastures (intensively managed 'high-producing exotic grassland', 25%), extensively managed 'low-producing grassland' (24%) and native tussock grassland (44%) [35]. Our measure of sheep/beef farming intensity was the percentage of each sub-catchment covered in 'high-producing exotic grassland', assuming that this land use would impose the strongest effects on nitrogen dynamics. Land cover data were available from the Landcover Database II [36] and a detailed delineation of stream reaches and their upstream catchments from the River Environmental Classification [37]. Overall, agricultural intensification in Central Otago has been relatively recent and land-use intensity is still increasing, therefore land-use legacy effects such as those reported elsewhere e.g., [38] are unlikely to be relevant in our case.

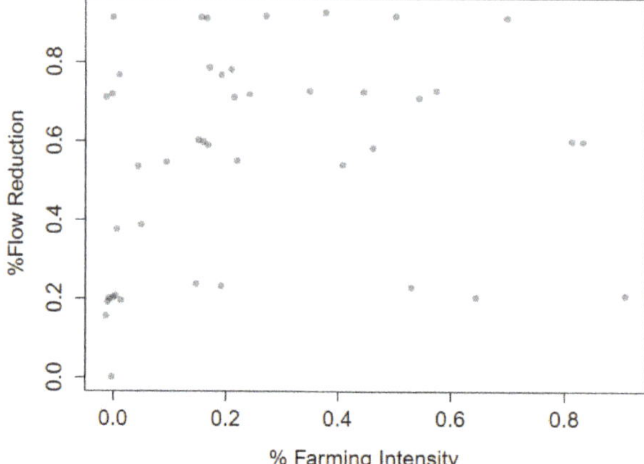

Figure 2. Distribution of the 43 stream sites along the gradients of % Farming Intensity and % Flow Reduction.

Flow reduction intensity was calculated from modelled data because the Manuherikia River catchment is subjected to five major water abstraction schemes, several dams and more than 238 individual water takes for which it is unknown exactly how much and when water is abstracted [39](Otago Regional Council, personal communication). Further, there is a complex network of water races transporting water within and between sub-catchments, making it almost impossible to estimate natural and current river flows for all reaches within the catchment. Moreover, it was not feasible to install flow gauges at all study sites. Kienzle and Schmidt [39] modelled streamflows for the Manuherikia catchment as runoff for hydrological response units based on rainfall, altitude, soil and vegetation properties while taking into account quickflow storage, groundwater storage and snow storage, using the ACRU model (Agricultural Catchments Research Unit; University of Natal, South Africa). They verified their model at four gauging stations with a high accuracy (ranging from 1.5 to 5.5%) for simulating the monthly totals. These four gauges were situated in headwater streams similar to our sites in size and geomorphology, but without any water abstraction. We therefore calculated flow reduction intensity as the percentage in streamflow reduction between the Dryland Scenario (current landcover but no water abstraction) and the Current Scenario by using the mean stream flow during five irrigation seasons from 1999/2000 to 2004/2005.

In the studied subcatchments, sheep/beef farming intensity ranged from 0 to 95% and flow reduction intensity from 0 to 92%. Invertebrates were sampled once from each site under base-flow conditions in Austral autumn, between 21 March and 4 April 2011.

2.2. Field Sampling and Sample Processing

Benthic stream invertebrates were sampled using a 500-µm mesh kick-net in pool and riffle habitats following standard methods for semi-quantitative collections [40]. We identified pools as deeper stream sections with slow, calm flow and riffles as shallower sections with relatively fast, turbulent flow. For each sample, we took invertebrates from 10 locations (depending on stream size from three to five different riffles or pools). Invertebrates were preserved in 70% ethanol and stored for two years prior to analysis. Although preservatives such as ethanol can affect δ^{15}N values, other studies have shown that the effects of ethanol on benthic macroinvertebrate stable isotope composition was minor and non-significant when compared to non-preserved material [41,42].

We focused on the most abundant grazer taxa in our study systems because we expected these to play a pivotal role in the local food webs. The three most common and widespread grazers at our sites were the snails *Potamopyrgus antipodarum* (present at 39 of 43 sites, comprising 21.1% of the total number of invertebrates counted and identified [34]) and *Physella acuta* (37 sites, 1.8%), and larvae of the mayfly *Deleatidium* spp. (21 sites with max. 53% farming intensity, 9.9%). For stable isotope analysis, we selected ten similar-sized individuals per taxon and sample (mean body length 3.8 mm for *Deleatidium*, mean shell lengths 2.3 mm for *Potamopyrgus* and 2.9 mm for *Physella*; measured to the nearest 0.1 mm under a dissecting microscope; Olympus SZ51, Olympus, Tokyo, Japan) and stored them in 90% ethanol in 5-mL glass vials. Snail shells and detritus were removed from individuals by soaking them in 1mL 1M HCl overnight and rinsing with deionised water. Gut contents were not removed due to the small body sizes and because of strong relationships between stable isotope values of primary consumer tissue and gut contents [43].

Grazers were dried at 60 °C for 48 h and subsampled (0.8 mg) into tin capsules. Nitrogen isotopes were measured by combusting all material to N_2 gas in an elemental analyser (Carlo Erba Instruments model NC2500, Milan, Italy). Gases were separated on a packed molecular sieve GC column and sequentially sent to an isotope ratio mass spectrometer ('20/20 Hydra', Europa Scientific, UK). Isotope values are reported as δ values (parts per thousand deviations from atmospheric N_2 standards):

$$\delta \text{ ‰} = \left(\frac{R_{sample}}{R_{standard}} - 1 \right) \times 1000 \tag{1}$$

where R is the heavy-to-light ratio of the isotope. A subset of samples was analyzed in duplicate and showed a mean standard deviation of 0.15‰.

2.3. Data Analysis

We investigated the relationships of $\delta^{15}N$ values of the three grazer taxa with sheep/beef farming intensity and flow reduction using general linear mixed models combined with information-theoretic model selection [44,45].

We examined 16 competing models for each grazer taxon including the global model (intercept plus five predictors: the first-order terms % Farming Intensity (FI) and % Flow Reduction (FR), habitat (riffle/pool), the second-order term FI × FI and the interaction FI × FR), simpler versions of the global model with one or more predictors removed, and the null model (intercept only). Sample sizes for each taxon were equivalent to the number of sites where the taxon occurred (39 sites for *P. antipodarum*, 37 for *P. acuta*, 21 for *Deleatidium*). All taxa were present at more than 16 sites, allowing us to examine all 16 models for each taxon. Site ID was included as a random effect to account for the spatial non-independence of the predictor habitat. The second-order term FI × FI was included to detect potential saturation-curve relationships such as reported by Larson et al. [8]. The second-order term FR × FR was not included because we had no hypothesis for this non-linear relationship. If the interaction or the second-order term were retained, the lower-order terms were also retained. The predictor 'habitat' was included because stable isotope values may differ between riffle and pool habitats due to different flow velocity conditions see e.g., [46]. All variables were centered by subtracting the sample mean (to improve interpretability of regression coefficients) and scaled with two standard deviations (to allow use of regression estimates as effect sizes) [47].

All models were ranked according to their AIC_c values (small sample unbiased Akaike Information Criterion) [48]. The top model set was chosen by selecting all models within $\Delta AIC_c \leq 2$ of the best model. Table 1 also shows the relative Akaike weights and marginal R^2 values (variance explained by the fixed effects) for the chosen top models. All analyses were computed in R (version 2.15) [49] using the packages lme4 [50] and MuMIn [51].

Table 1. Standardized partial regression estimates (effect sizes, ES), 95% confidence intervals CIs, delta AIC (Akaike Information Criterion) differences to the model ranked most highly (ΔAICc), Akaike weights (re-scaled), marginal R^2 values (variation explained by the fixed effects) of the final models for the relationships of δ^{15}N with sheep/beef farming intensity and flow reduction. Effect size categories after Nakagawa & Cuthill [52]: weak > 0.10, moderate > 0.30, strong > 0.50. Estimates in bold indicate where predictors had a significant effect on the response variable (95% CIs not including zero). FI = % Farming Intensity, FR = % Flow Reduction, FI × FI = second-order polynomial terms and FI × FR = interaction terms.

Response Variable	Intercept	95% CIs	FI ES	95% CIs	FI × FI ES	95% CIs	FR ES	95% CIs	FI × FR ES	95% CIs	ΔAIC$_c$	Weight	Marginal R^2
δ^{15}N *Deleatidium*	0.12	(−0.02, 0.26)	**1.23**	(0.86, 1.60)	**−0.48**	(−0.83, −0.12)					0.00	0.51	0.75
	0.14	(0.01, 0.29)	**0.94**	(0.59, 1.29)			0.15	(−0.14, 0.44)	**−0.92**	(−1.70, −0.14)	1.48	0.25	0.76
	0.01	(−0.12, 0.15)	**0.84**	(0.56, 1.13)							1.53	0.24	0.66
δ^{15}N *Potamopyrgus*	−0.01	(−0.19, 0.16)									0.00	0.73	0.00
	−0.01	(−0.19, 0.16)					0.24	(−0.09, 0.57)			1.95	0.27	0.06
δ^{15}N *Physella*	−0.02	(−0.19, 0.15)									0.00	0.52	0.00
	−0.02	(−0.18, 0.14)					0.26	(−0.06, 0.57)			1.57	0.24	0.06
	−0.03	(−0.19, 0.14)	−0.26	(−0.58, 0.06)							1.58	0.24	0.06

3. Results

Primary consumer stable isotope values spanned a wide range (*Deleatidium* δ^{15}N: 1–9.9‰, *Potamopyrgus* δ^{15}N: 1.8–13.2‰, *Physella* δ^{15}N: 4.3–14.7‰). Model selections results revealed inconsistent relationships for the δ^{15}N values of the three grazers with farming intensity and flow reduction. The δ^{15}N values of the two snails were more similar than the δ^{15}N values of *Deleatidium* compared to *Potamopyrgus* and *Physella* (Figure 3).

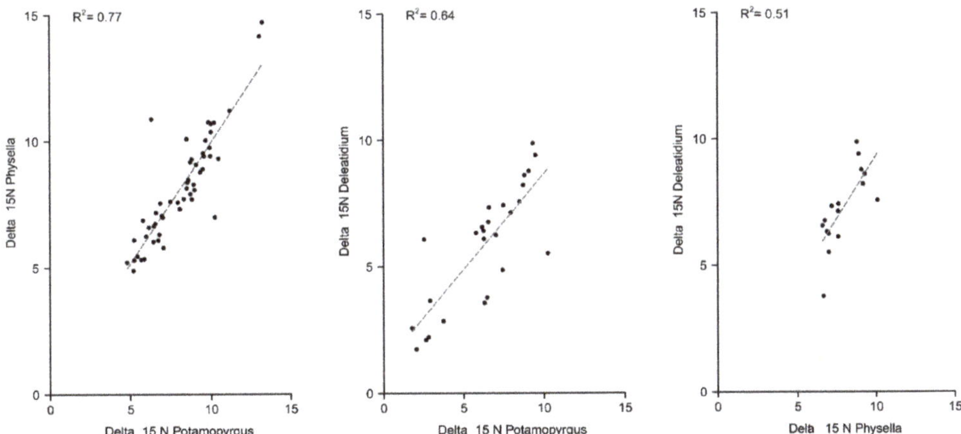

Figure 3. Comparison of primary consumers δ^{15}N ratios at stream sites where more than one grazer taxon occurred. The fitted linear regressions between taxa δ^{15}N values are represented by the dashed lines and the strength of the relationships by the R^2 values.

Deleatidium δ^{15}N values were best modelled by the single-stressor model including the first-order and second-order terms for farming intensity, explaining 75% of the variation (curvilinear relationship, best described as a saturation curve; Table 1, Figure 4a). The second-best model for *Deleatidium* δ^{15}N values was a complex multiple-stressor model, in which the relationship with one stressor gradient depended on the intensity of the second stressor. This model explained 76% of the variation in the data, with strong positive effect sizes for farming intensity and the interaction term, plus a weak positive effect size for flow reduction (Figure 4b). The best models for *Potamopyrgus* and *Physella* δ^{15}N values indicated no relationship with farming intensity or flow reduction (null model; Table 1). The second-best models for *Potamopyrgus* (Figure 4d) and *Physella* δ^{15}N values (Figure 4e) both included a positive but non-significant relationship with flow reduction (both models only explained 6% of the variation).

The interaction term in the second-best model for *Deleatidium* δ^{15}N indicated an antagonistic interaction where the relationship with farming intensity was weaker at high flow reduction and the relationship with flow reduction was also weaker at high farming intensity (Figure 4b).

The predictor 'habitat type' (riffle versus pool) was not retained in any of the final models.

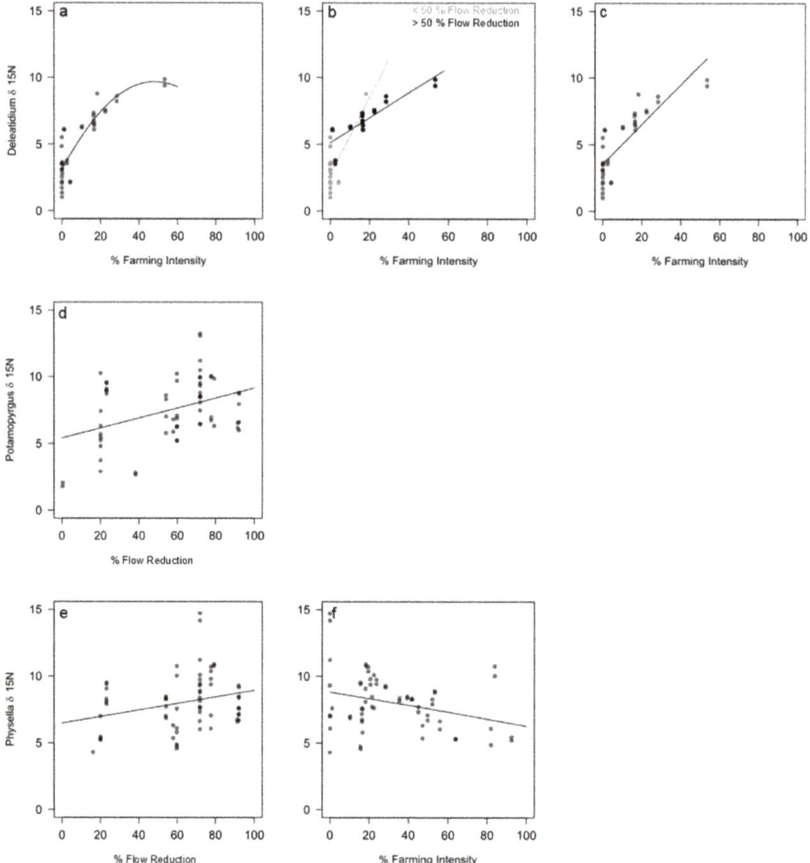

Figure 4. Relationships between primary consumer δ^{15}N values and sheep/beef farming intensity and flow reduction derived from the model selection procedure. The plots show the relationships for each of the competing best models including farming intensity or flow reduction as predictor variables (3 models for *Deleatidium* (**a–c**), 1 for *Potamopyrgus* (**d**) and 2 for *Physella* (**e,f**); see Table 1 for details). The plots encompass the range of stressor values across the study sites and riffle/pool habitats (note the plotted range of farming intensities is smaller for *Deleatidium*). The interaction plot for *Deleatidium* (**b**) shows that the relationship with % Farming Intensity is more positive at lower % Flow Reduction (for ease of interpretation, % Flow Reduction was labelled as a factor with 2 levels; grey < 50% Flow Reduction, black > 50% Flow Reduction).

4. Discussion

4.1. Do Primary Consumers Show Similar Relationships along Stressor Gradients

Ecologists have assumed that primary consumers mirror the stable isotope values of their periphyton food sources [32,53], and stable isotope values of different primary consumer taxa have been used interchangeably to assess anthropogenic disturbances in some previous freshwater studies [7,8]. In contrast, our three consumer taxa followed different response patterns in terms of direction and strength of observed relationships along our focal gradients of stressor intensity (see discussion below). Thus, we reject our hypothesis 1 that primary consumers show similar stable isotope patterns along stressor gradients.

Deleatidium δ^{15}N values showed a strong relationship with sheep/beef farming intensity, whereas the best models for *Potamopyrgus* and *Physella* δ^{15}N values indicated that this predictor was not able to explain a sufficient amount of variation in the data. Streamflow reduction only featured in the second-best models for *Deleatidium* and the two snails. The tighter relationship for *Deleatidium* may partly arise because of its absence from sites with more than 53% farming intensity that may be subject to high nitrogen concentrations and sediment levels [34]. Such physicochemical conditions may provide unsuitable habitat for this mayfly larva, which cannot feed on thick layers of filamentous green algae and biofilm attached to fine sediment [54,55], but they may suit snails such as *Potamopyrgus* that can exploit filamentous greens as well as epipsammic biofilms [54,56]. Thus, *Deleatidium* with its sweep-like mouthparts feeds preferentially on erect, tall-growing benthic diatoms (epilithon) and long filamentous algae are rarely ingested [57,58]. By contrast, *Potamopyrgus* possesses a radula that can scrape off tightly attached biofilms, enabling them to remove biofilm from surface and subsurface sediments [54] and even from fine or coarse particulate organic matter [59]. It is likely that *Potamopyrgus* can access a larger range of food resources than *Deleatidium*, helping to explain the snail's success in so many different habitats and as an invader [57]. The snail *Physella* also has a radula, therefore its primary feeding mode is likely to be similar to *Potamopyrgus*. In support of this theory, the relationship between the δ^{15}N values of the two snails from the same sites was tighter than their respective relationships with *Deleatidium* (Figure 3). Other possible explanations for differences in δ^{15}N values between *Deleatidium* and the two snails might relate to their respective physiology, mobility and life history.

A limitation of our study is the lack of data on consumer diet composition and the δ^{15}N values of all potential diet sources. However, given that we collected invertebrate consumers from the same habitats, differences in consumer δ^{15}N values were probably related to ingestion of different components of the periphytic community. Potential food sources such as diatoms, filamentous green algae and leaf material differed in their isotopic values in a pasture stream in Waikato, another region of New Zealand dominated by farmland [56].

4.2. Multiple-Stressor Patterns of Primary Consumer Stable Isotope Values

Although survey-based studies such as ours are less suitable for explaining cause-and-effect relationships than manipulative experiments [60], the observed relationships between stable isotope values and our catchment-scale variables can be regarded as robust for one of our three response variables, *Deleatidium*, for which we were able to explain up to 76% of the variation.

At the catchment scale, *Deleatidium* δ^{15}N values were positively related to rising sheep/beef farming intensity and also to increasing flow reduction, but δ^{15}N values of the snails were either not or only weakly related to rising farming intensity (only partially supporting hypothesis 2). For δ^{15}N values of *Deleatidium*, which showed by far the strongest patterns of the three consumers (but only occurred at 21 of the 43 sites and only in streams with up to 53.4% farming intensity), the best model was an antagonistic one in which the response pattern along one stressor gradient depended on the intensity of the second stressor (in support of hypothesis 3).

4.3. Differences in Consumers' Relationships with Sheep/Beef Farming Intensity

Four processes may have contributed to the increase in δ^{15}N values in *Deleatidium* with rising sheep/beef farming intensity: inputs of industrial fertilizer (initial δ^{15}N values between +3‰ and −3‰; mean around 0‰), inputs of animal waste products (δ^{15}N values from +35‰ to −15‰; mean around +10‰), nitrogen transformation processes such as denitrification and ammonia volatilization in agricultural soils and streams [18,61], and atmospheric nitrogen deposition. The latter probably plays a lesser role in New Zealand than in many other developed countries, but all three other processes likely contributed to the observed δ^{15}N patterns. High-producing exotic grasslands used for sheep/beef farming in New Zealand receive not just animal manure, but also regular inputs of industrial fertilizers, via a process called "topdressing" (often done using small light airplanes). Further, field observations of the streambed substrata during invertebrate sampling, combined with a strong sulfur smell at some study sites (K. Lange, pers. comm.), suggested that anoxic conditions in the sediment, which increase denitrification rates in a zone of very low oxygen conditions [17], may have occurred at some of the 43 sites. Because our study was observational rather than manipulative and we had no information on manure or fertilizer inputs from the stream catchments, we cannot determine the relative importance of these three processes in determining the patterns in our δ^{15}N data.

The curvilinear relationship of *Deleatidium* δ^{15}N values with farming intensity in our study contrasts with the positive linear relationships reported in previous studies in New Zealand [7] and elsewhere [6]. In these other studies, Anderson and Cabana [6] focused on nitrogen load from manure while Clapcott et al. [7] studied agriculture, forestry and urban areas, contrasting with our investigation of the percentage of high-intensity farming in an agricultural catchment. Anderson and Cabana [6] considered agricultural prevalence as a land-use intensity gradient, but their gradient only extended to a maximum of 52% of the catchment area. The possibility that manure from different animals may also differ in isotopic composition could also add to the observed variation between studies. Perhaps we were able to detect a positive relationship approximating a saturation curve because our sites covered a wider gradient of land-use intensity than Anderson and Cabana [6]. Larson et al. [8] also reported a non-linear relationship between δ^{15}N and catchment land-use intensity (% agriculture) that was best explained by a saturation curve, and suggested the non-linearity may have been caused by biotic nitrogen accumulation itself being non-linear. Under such circumstances, phosphorus might have been the limiting factor for primary productivity so that further nitrogen inputs could not be utilized and turned into biomass. It is also possible that the bioavailability of nitrogen differed among sources such as inorganic fertilizer, manure and humic substances.

Potamopyrgus showed no relationship with farming intensity and we speculate that these snails may have switched to other resources where tightly-attached epilithic algae were unavailable or where other resources were more common, such as filamentous green algae, macrophytes, detritus or heterotrophic biofilms, all of which likely had δ^{15}N values differing from that of tightly-attached epilithic algae [61].

The weak negative relationship of *Physella* δ^{15}N values with farming intensity was unexpected. One possible explanation for the contrasting relationships of *Physella* and *Deleatidium* δ^{15}N values is exploitation of different microhabitats. The snail tended to be associated with streams having intermediate farming intensities, whereas *Deleatidium* abundances were associated with lower farming intensities [34]. Therefore, it was unsurprising that the taxa were sympatric in only a few sites (Figure 3). These sites may have provided a variety of habitats including rocky surfaces for *Deleatidium* and macrophytes for *Physella*. Future studies that would determine stable isotope composition of all basal resources at a given site would permit gaining a better understanding of the consumer's resource use, as well as assessment of microhabitat use.

4.4. Positive Consumer Relationships with Flow Reduction Intensity

The positive relationships of δ^{15}N values with flow reduction intensity in our study could be due to streams subjected to high flow reduction experiencing higher denitrification rates because of reduced flow velocities. Denitrification in streams mainly occurs in oxygen-depleted zones of bed sediments, and denitrification rates can be increased by high rates of respiration and anoxic conditions following excess rates of primary production related to increased water retention times [17]. Alternatively, weaker dilution of the concentrations of nitrogen from anthropogenic sources due to reduced stream flows could also result in increased δ^{15}N values in the tissues of grazers.

For *Deleatidium*, the intensity of sheep/beef farming was more important in determining stable isotope values than hydrological alteration. In a companion study of biological invertebrate traits in the same river catchment [35], we also found that farming intensity had stronger effects on habitat availability and flows of matter and energy than flow reduction. It is likely that agriculture imposes more direct stressors on stream environments (e.g., sediment and nutrient inputs, reduced shading) than hydrological alteration (e.g., increased retention time and reduced peak flows). On the other hand, our choice of study sites spanning 3rd to 5th order streams may have included some differences in flow regimes between individual sites, and these may have contributed to the overall variation in our non-manipulative study, weakening the observed flow reduction effects.

4.5. Antagonistic Stressor Interaction

The interaction term between sheep/beef farming intensity and flow reduction was retained in one of the final models for *Deleatidium* $\delta^{15}N$ values. This interaction was classified as an antagonism because the combined multiple-stressor effects on stable isotope values were weaker than one would have expected based on their respective individual effects. In general, it is difficult to find evidence for synergisms if both stressors have large individual effects and the sum of the individual effects exceeds 100% [3]. In our study, this was the case for *Deleatidium* where individual stressors effects already caused a strong increase in $\delta^{15}N$ values. Moreover, antagonisms can also pose significant management challenges since the intensities of all interacting stressors may need to be reduced to achieve substantial recovery [62].

The interaction between farming intensity and flow reduction for *Deleatidium* could be interpreted as follows: Overall, $\delta^{15}N$ values of *Deleatidium* increased with farming intensity (the strongest effect in the model) due to reasons discussed earlier on. Low levels of flow reduction appear to have little effect on this positive relationship. At high levels of flow reduction, however, nitrogen inputs from adjoining pastures via the flowing water might be decreased to such an extent that the positive effect of increased farming intensity on $\delta^{15}N$ values is weakened.

5. Conclusions

Based on our findings, we believe that the mayfly *Deleatidium* may be well-suited as a bioindicator in stable isotope studies on agricultural impacts because its $\delta^{15}N$ values were strongly related to sheep/beef farming intensity. Moreover, the mayfly provides a time-integrated measure (1–2 generations per year) [63,64] and occurs throughout New Zealand [57,65]. Further, sampling of this mayfly in the field is easy because it often reaches high densities, and the techniques for stable isotope analysis are well established. Interestingly, in terms of explaining variation in our stream survey data, $\delta^{15}N$ values of *Deleatidium* ($R^2 = 0.76$) performed better than structural measures of invertebrate community composition (e.g., taxonomic richness of pollution-sensitive mayflies, stoneflies and caddis flies, $R^2 = 0.42$) [34] when detecting impacts at the catchment scale.

In stream surveys or experiments focusing on land-use effects, the same primary consumers are unlikely to be present at all study sites or in all experimental units, especially if these sites/units span broad gradients of stressor intensities. We caution against extrapolating 'missing' values for one taxon with data from other taxa (as done in [7] and [8]). Instead we recommend establishing baseline values and closely investigating resource use and multiple-stressor relationships for several primary consumers, ideally focusing on taxa with strict dietary preferences, such as *Deleatidium*.

Finally, we observed one complex interaction between paired stressors (out of three possible cases) in our analysis. Such non-additive response patterns have also been found for other functional or structural metrics in freshwater [60,66], terrestrial [67,68] and marine environments [62,69]. Thus, our findings suggest that non-linear and non-additive responses of consumer stable isotope values to multiple stressors may be fairly common and should therefore be considered in future studies.

Author Contributions: Conceptualization, K.L.; Methodology, K.L., C.R.T. and C.D.M.; Software, K.L. and C.D.M.; Validation, K.L. and C.D.M.; Formal Analysis, K.L. and C.D.M.; Investigation, K.L., C.R.T. and C.D.M.; Resources, K.L., C.R.T. and C.D.M.; Data Curation, K.L.; Writing—Original Draft Preparation, K.L.; Writing—Review & Editing, K.L., C.R.T. and C.D.M.; Visualization, K.L., C.R.T. and C.D.M.; Supervision, C.R.T. and C.D.M.; Funding Acquisition, K.L., C.R.T. and C.D.M.

Funding: This research was funded by the NZ Ministry of Business, Innovation & Employment (grant number C01 × 1005). K.L. further acknowledges funding through a University of Otago Postgraduate Scholarship and the Department of Zoology.

Acknowledgments: We thank Jürgen Kienzle and Stefan Schmidt for providing modelled hydrological data; Pierre Chanut and Fabiana Schneck for assistance with fieldwork; Nicky McHugh and Dianne Clark for help in the laboratory; Bob L. de Berry for accommodation during fieldwork and the Otago Freshwater Ecology Group for discussions.

Conflicts of Interest: The authors declare no conflict of interest.

References

1. Vinebrooke, R.D.; Cottingham, K.L.; Norberg, M.S.J.; Dodson, S.I.; Maberly, S.C.; Sommer, U.; Norberg, M.S. Impacts of multiple stressors on biodiversity and ecosystem functioning: The role of species co-tolerance. *Oikos* **2004**, *104*, 451–457. [CrossRef]

2. Ormerod, S.J.; Dobson, M.; Hildrew, A.G.; Townsend, C.R. Multiple stressors in freshwater ecosystems. *Freshw. Boil.* **2010**, *55*, 1–4. [CrossRef]

3. Folt, C.L.; Chen, C.Y.; Moore, M.V.; Burnaford, J. Synergism and antagonism among multiple stressors. *Limnol. Oceanogr.* **1999**, *44*, 864–877. [CrossRef]

4. Vivian, C. Tracers of sewage sludge in the marine environment: A review. *Sci. Total Environ.* **1986**, *53*, 5–40. [CrossRef]

5. Peterson, B.J.; Fry, B. Stable Isotopes in Ecosystem Studies. *Annu. Rev. Ecol. Syst.* **1987**, *18*, 293–320. [CrossRef]

6. Anderson, C.; Cabana, G. Does δ^{15}N in river food webs reflect the intensity and origin of N loads from the watershed? *Sci. Total Environ.* **2006**, *367*, 968–978. [CrossRef]

7. Clapcott, J.E.; Young, R.G.; Goodwin, E.O.; Leathwick, J.R. APPLIED ISSUES: Exploring the response of functional indicators of stream health to land-use gradients. *Freshw. Boil.* **2010**, *55*, 2181–2199. [CrossRef]

8. Larson, J.H.; Richardson, W.B.; Vallazza, J.M.; Nelson, J.C. Rivermouth Alteration of Agricultural Impacts on Consumer Tissue δ^{15}N. *PLoS ONE* **2013**, *8*, e69313. [CrossRef]

9. Lake, J.L.; McKinney, R.A.; Osterman, F.A.; Pruell, R.J.; Kiddon, J.; Ryba, S.A.; Libby, A.D. Stable nitrogen isotopes as indicators of anthropogenic activities in small freshwater systems. *Can. J. Fish. Aquat. Sci.* **2001**, *58*, 870–878. [CrossRef]

10. Cabana, G.; Rasmussen, J.B. Comparison of aquatic food chains using nitrogen isotopes. *Proc. Natl. Acad. Sci. USA* **1996**, *93*, 10844–10847. [CrossRef]

11. McClelland, J.W.; Valiela, I. Linking nitrogen in estuarine producers to land-derived sources. *Limnol. Oceanogr.* **1998**, *43*, 577–585. [CrossRef]

12. Vermeulen, S.; Sturaro, N.; Gobert, S.; Bouquegneau, J.; Lepoint, G. Potential early indicators of anthropogenically derived nutrients: A multiscale stable isotope analysis. *Mar. Ecol. Prog. Ser.* **2011**, *422*, 9–22. [CrossRef]

13. Barr, N.G.; Dudley, B.D.; Rogers, K.M.; Cornelisen, C.D. Broad-scale patterns of tissue-δ^{15}N and tissue-N indices in frondose Ulva spp.; developing a national baseline indicator of nitrogen-loading for coastal New Zealand. *Mar. Pollut. Bull.* **2013**, *67*, 203–216. [CrossRef] [PubMed]

14. Allan, J.D. Landscapes and Riverscapes: The Influence of Land Use on Stream Ecosystems. *Annu. Rev. Ecol. Evol. Syst.* **2004**, *35*, 257–284. [CrossRef]

15. Davies-Colley, R.J.; Meleason, M.A.; Hall, R.M.; Rutherford, J.C. Modelling the time course of shade, temperature, and wood recovery in streams with riparian forest restoration. *N. Z. J. Mar. Freshw. Res.* **2009**, *43*, 673–688. [CrossRef]

16. Dewson, Z.S.; James, A.B.W.; Death, R.G. A review of the consequences of decreased flow for instream habitat and macroinvertebrates. *J. N. Am. Benthol. Soc.* **2007**, *26*, 401–415. [CrossRef]

17. Seitzinger, S.; Harrison, J.A.; Bohlke, J.K.; Bouwman, A.F.; Lowrance, R.; Peterson, B.; Tobias, C.; Van Drecht, G. Denitrification across landscapes and waterscapes: A synthesis. *Ecol. Appl.* **2006**, *16*, 2064–2090. [CrossRef]

18. Anderson, C.; Cabana, G. δ^{15}N in riverine food webs: Effects of N inputs from agricultural watersheds. *Can. J. Fish. Aquat. Sci.* **2005**, *62*, 333–340. [CrossRef]

19. Mayer, B.; Boyer, E.W.; Goodale, C.; Jaworski, N.A.; Van Breemen, N.; Howarth, R.W.; Seitzinger, S.; Billen, G.; Lajtha, K.; Nadelhoffer, K.; et al. Sources of nitrate in rivers draining sixteen watersheds in the northeastern U.S.: Isotopic constraints. *Biogeochemistry* **2002**, *57*, 171–197. [CrossRef]

20. Sébilo, M.; Billen, G.; Grably, M.; Mariotti, A. Isotopic composition of nitrate-nitrogen as a marker of riparian and benthic denitrification at the scale of the whole Seine River system. *Biogeochemistry* **2003**, *63*, 35–51. [CrossRef]

21. Kellman, L.; Hillaire-Marcel, C. Nitrate cycling in streams: Using natural abundances of $NO_3^- $-$\delta^{15}$N to measure in-situ denitrification. *Biogeochemistry* **1998**, *43*, 273–292. [CrossRef]

22. Diebel, M.W.; Zanden, M.J.V. Nitrogen stable isotopes in streams: Effects of agricultural sources and transformations. *Ecol. Appl.* **2009**, *19*, 1127–1134. [CrossRef] [PubMed]

23. Kendall, C. Tracing Nitrogen Sources and Cycling in Catchments. In *Isotope Tracers in Catchment Hydrology*; Kendall, C., McDonnell, J.J., Eds.; Elsevier: Amsterdam, The Netherlands, 1998; pp. 519–576.

24. Voss, M.; Deutsch, B.; Elmgren, R.; Humborg, C.; Kuuppo, P.; Pastuszak, M.; Rolff, C.; Schulte, U. Source identification of nitrate by means of isotopic tracers in the Baltic Sea catchments. *Biogeosciences* **2006**, *3*, 663–676. [CrossRef]

25. Barnes, R.T.; Raymond, P.A. Land-use controls on sources and processing of nitrate in small watersheds: Insights from dual isotopic analysis. *Ecol. Appl.* **2010**, *20*, 1961–1978. [CrossRef]

26. Udy, J.W.; Fellows, C.S.; Bartkow, M.E.; Bunn, S.E.; Clapcott, J.E.; Harch, B.D. Measures of Nutrient Processes as Indicators of Stream Ecosystem Health. *Hydrobiologia* **2006**, *572*, 89–102. [CrossRef]

27. Udy, J.W.; Bunn, S.E. Elevated delta N-15 values in aquatic plants from cleared catchments: Why? *Mar. Freshw. Res.* **2001**, *52*, 347–351. [CrossRef]

28. Atkinson, C.L.; Christian, A.D.; Spooner, D.E.; Vaughn, C.C. Long-lived organisms provide an integrative footprint of agricultural land use. *Ecol. Appl.* **2014**, *24*, 375–384. [CrossRef]

29. Anderson, C.; Cabana, G. Estimating the trophic position of aquatic consumers in river food webs using stable nitrogen isotopes. *J. N. Am. Benthol. Soc.* **2007**, *26*, 273–285. [CrossRef]

30. Finlay, J.C. Patterns and controls of lotic algal stable carbon isotope ratios. *Limnol. Oceanogr.* **2004**, *49*, 850–861. [CrossRef]

31. Jardine, T.D.; Kidd, K.A.; Fisk, A.T. Applications, Considerations, and Sources of Uncertainty When Using Stable Isotope Analysis in Ecotoxicology. *Environ. Sci. Technol.* **2006**, *40*, 7501–7511. [CrossRef]

32. Jardine, T.D.; Hadwen, W.L.; Hamilton, S.K.; Hladyz, S.; Mitrovic, S.M.; Kidd, K.A.; Tsoi, W.Y.; Spears, M.; Westhorpe, D.P.; Fry, V.M.; et al. Understanding and covercoming baseline isotopic variability in running waters. *River Res. Appl.* **2014**, *30*, 155–165. [CrossRef]

33. Jackson, M.C.; Loewen, C.J.G.; Vinebrooke, R.D.; Chimimba, C.T. Net effects of multiple stressors in freshwater ecosystems: A meta-analysis. *Glob. Chang. Biol.* **2016**, *22*, 180–189. [CrossRef] [PubMed]

34. Lange, K.; Townsend, C.R.; Matthaei, C.D. Can biological traits of stream invertebrates help disentangle the effects of multiple stressors in an agricultural catchment? *Freshw. Boil.* **2014**, *59*, 2431–2446. [CrossRef]

35. Lange, K.; Townsend, C.R.; Gabrielsson, R.; Chanut, P.C.M.; Matthaei, C.D. Responses of stream fish populations to farming intensity and water abstraction in an agricultural catchment. *Freshwat. Biol.* **2014**, *59*, 286–299. [CrossRef]

36. Ministry for the Environment. *Land Cover Database II (LCDB2)*; Ministry for the Environment: Wellington, New Zealand, 2008.

37. Ministry for the Environment. *River Environmental Classification (REC)*; Ministry for the Environment: Wellington, New Zealand, 2010.

38. Harding, J.S.; Benfield, E.F.; Bolstad, P.V.; Helfman, G.S.; Jones, E.B.D. Stream biodiversity: The ghost of land use past. *Proc. Natl. Acad. Sci. USA* **1998**, *95*, 14843–14847. [CrossRef]

39. Kienzle, S.W.; Schmidt, J. Hydrological impacts of irrigated agriculture in the Manuherikia catchment, Otago, New Zealand. *J. Hydrol. N. Z.* **2008**, *47*, 67–84.

40. Stark, J.D.; Boothroyd, I.K.G.; Harding, J.S.; Maxted, J.R.; Scarsbrook, M.R. *Protocols for Sampling Macroinvertebrates in Wadeable Streams*; Ministry for the Environment: Wellington, New Zealand, 2001; p. 48.

41. Syväranta, J.; Vesala, S.; Rask, M.; Ruuhijärvi, J.; Jones, R. Evaluating the utility of stable isotope analyses of archived freshwater sample materials. *Hydrobiologia* **2008**, *600*, 121–130. [CrossRef]

42. Lau, D.C.P.; Leung, K.M.Y.; Dudgeon, D. Preservation effects on C/N ratios and stable isotope signatures of freshwater fishes and benthic macroinvertebrates. *Limnol. Oceanogr. Methods* **2012**, *10*, 75–89. [CrossRef]

43. Jardine, T.D.; Curry, R.A.; Heard, K.S.; Cunjak, R.A. High fidelity: Isotopic relationship between stream invertebrates and their gut contents. *J. N. Am. Benthol. Soc.* **2005**, *24*, 290–299. [CrossRef]

44. Johnson, J.B.; Omland, K.S. Model selection in ecology and evolution. *Trends Ecol. Evol.* **2004**, *19*, 101–108. [CrossRef]

45. Wagenhoff, A.; Townsend, C.R.; Phillips, N.; Matthaei, C.D. Subsidy-stress and multiple-stressor effects along gradients of deposited fine sediment and dissolved nutrients in a regional set of streams and rivers. *Freshw. Boil.* **2011**, *56*, 1916–1936. [CrossRef]

46. Finlay, J.C.; Power, M.E.; Cabana, G. Effects of water velocity on algal carbon isotope ratios: Implications for river food web studies. *Limnol. Oceanogr.* **1999**, *44*, 1198–1203. [CrossRef]

47. Schielzeth, H. Simple means to improve the interpretability of regression coefficients. *Methods Ecol. Evol.* **2010**, *1*, 103–113. [CrossRef]

48. Burnham, K.P.; Anderson, D.R. *Model Selection and Multimodel Inference: A Practical Information-Theoretic Approach*; Springer: New York, NY, USA, 2002; Volume 2, p. 488.

49. R Development Core Team. *R: A Language and Environment for Statistical Computing, 3.0.2*; R Development Core Team: Vienna, Austria, 2014.

50. Bates, D.; Maechler, M.; Bolker, B.; Walker, S. *Lme4: Mixed-Effects Modeling with R, 1.1-7*. 2015. Available online: https://mran.microsoft.com/snapshot/2016-03-04/web/packages/lme4/README.html (accessed on 26 October 2019).

51. Bartoń, K. *MuMIn: Multi-Model Inference, 1.9.0*. 2013. Available online: http://www2.uaem.mx/r-mirror/web/packages/MuMIn/MuMIn.pdf (accessed on 26 October 2019).

52. Nakagawa, S.; Cuthill, I.C. Effect size, confidence interval and statistical significance: A practical guide for biologists. *Boil. Rev.* **2007**, *82*, 591–605. [CrossRef] [PubMed]

53. Post, D.M. Using stable isotopes to estimate trophic position: Models, methods, and assumptions. *Ecology* **2002**, *83*, 703–718. [CrossRef]

54. Broekhuizen, N.; Parkyn, S.; Miller, D. Fine sediment effects on feeding and growth in the invertebrate grazers *Potamopyrgus antipodarum* (Gastropoda, Hydrobiidae) and *Deleatidium* sp. (Ephemeroptera, Leptophlebiidae). *Hydrobiologia* **2001**, *457*, 125–132. [CrossRef]

55. Finlay, J.C. Stable-Carbon-Isotope Ratios of River Biota: Implications for Energy Flow in Lotic Food Webs. *Ecology* **2001**, *82*, 1052. [CrossRef]

56. Hicks, B.J. Food webs in forest and pasture streams in the Waikato region, New Zealand: A study based on analyses of stable isotopes of carbon and nitrogen, and fish gut contents. *N. Z. J. Mar. Freshw. Res.* **1997**, *31*, 651–664. [CrossRef]

57. Holomuzki, J.R.; Biggs, B.J.F. Food Limitation Affects Algivory and Grazer Performance for New Zealand Stream Macroinvertebrates. *Hydrobiologia* **2006**, *561*, 83–94. [CrossRef]

58. Rounick, J.S.; Winterbourn, M.J. The formation, structure and utilization of stone surface organic layers in two New Zealand streams. *Freshw. Boil.* **1983**, *13*, 57–72. [CrossRef]

59. Parkyn, S.M.; Quinn, J.M.; Cox, T.J.; Broekhuizen, N. Pathways of N and C uptake and transfer in stream food webs: An isotope enrichment experiment. *J. N. Am. Benthol. Soc.* **2005**, *24*, 955–975. [CrossRef]

60. Townsend, C.R.; Uhlmann, S.S.; Matthaei, C.D. Individual and combined responses of stream ecosystems to multiple stressors. *J. Appl. Ecol.* **2008**, *45*, 1810–1819. [CrossRef]

61. Finlay, J.C.; Kendall, C. Stable isotope tracing of temporal and spatial variability in organic matter sources to freshwater ecosystems. In *Stable Isotopes in Ecology and Environmental Science*; Michener, R.H., Lajtha, K., Eds.; Blackwell Publishing: Singapore, 2007; pp. 283–333.

62. Crain, C.M.; Kroeker, K.; Halpern, B.S. Interactive and cumulative effects of multiple human stressors in marine systems. *Ecol. Lett.* **2008**, *11*, 1304–1315. [CrossRef] [PubMed]

63. Huryn, A.D. Temperature-dependent growth and life cycle of Deleatidium (Ephemeroptera: Leptophlebiidae) in two high-country streams in New Zealand. *Freshwat. Biol.* **1996**, *36*, 351–361. [CrossRef]

64. Dybdahl, M.F.; Kane, S.L. Adaptation vs. phenotypic plasticity in the success of a clonal invader. *Ecology* **2005**, *86*, 1592–1601. [CrossRef]

65. Broekhuizen, N.; Parkyn, S.; Miller, D.; Rose, R. The relationship between food density and short term assimilation rates in *Potamopyrgus antipodarum* and *Deleatidium* sp. *Hydrobiologia* **2002**, *477*, 181–188. [CrossRef]

66. Piggott, J.J.; Niyogi, D.K.; Townsend, C.R.; Matthaei, C.D. Multiple stressors and stream ecosystem functioning: Climate warming and agricultural stressors interact to affect processing of organic matter. *J. Appl. Ecol.* **2015**, *52*, 1126–1134. [CrossRef]

67. Aebi, A.; Neumann, P. Endosymbionts and honey bee colony losses? *Trends Ecol. Evol.* **2011**, *26*, 494. [CrossRef]
68. Potts, S.G.; Biesmeijer, J.C.; Kremen, C.; Neumann, P.; Schweiger, O.; Kunin, W.E. Global pollinator declines: Trends, impacts and drivers. *Trends Ecol. Evol.* **2010**, *25*, 345–353. [CrossRef]
69. McLeod, E.; Anthony, K.R.; Andersson, A.; Beeden, R.; Golbuu, Y.; Kleypas, J.; Kroeker, K.; Manzello, D.; Salm, R.V.; Schuttenberg, H.; et al. Preparing to manage coral reefs for ocean acidification: Lessons from coral bleaching. *Front. Ecol. Environ.* **2013**, *11*, 20–27. [CrossRef]

Article

Multiple-Line Identification of Socio-Ecological Stressors Affecting Aquatic Ecosystems in Semi-Arid Countries: Implications for Sustainable Management of Fisheries in Sub-Saharan Africa

Vincent-Paul Sanon [1,2], Patrice Toé [2], Jaime Caballer Revenga [1], Hamid El Bilali [3], Laura Janine Hundscheid [1], Michalina Kulakowska [4], Piotr Magnuszewski [4,5], Paul Meulenbroek [6], Julie Paillaugue [1], Jan Sendzimir [1,6], Gabriele Slezak [7], Stefan Vogel [8] and Andreas H. Melcher [1,*]

[1] Institute for Development Research, BOKU—University of Natural Resources and Life Sciences, Peter-Jordan-Strasse 76/I, 1190 Vienna, Austria; vincent.sanon@boku.ac.at (V.-P.S.); jacabenga@gmail.com (J.C.R.); laura.hundscheid@boku.ac.at (L.J.H.); j.paillaugue@lilo.org (J.P.); jan.sendzimir@boku.ac.at (J.S.)
[2] Institute for Rural Development, University Nazi BONI, Bobo-Dioulasso 01 BP 1091 Bobo 01, Burkina Faso; patrice_toe57@yahoo.fr
[3] International Centre for Advanced Mediterranean Agronomic Studies, via Ceglie 9, 70010 Valenzano (Bari), Italy; elbilali@iamb.it
[4] Centre for Systems Solutions, Stefana Jaracza 80b/10, 50-305 Wrocław, Poland; michalina.kulakowska@crs.org.pl (M.K.); piotr.magnuszewski@crs.org.pl (P.M.)
[5] International Institute for Applied Systems Analysis, Schlossplatz 1, A-2361 Laxenburg, Austria
[6] Institute of Hydrobiology and Aquatic Ecosystem Management, BOKU—University of Natural Resources and Life Sciences, Gregor-Mendel-Strasse 33, 1180 Vienna, Austria; paul.meulenbroek@boku.ac.at
[7] Institute for African Studies, University of Vienna, Spitalgasse 2, 1090 Vienna, Austria; gabriele.slezak@univie.ac.at
[8] Institute for Sustainable Economic Development, BOKU—University of Natural Resources and Life Sciences, Feistmantelstraße 4, 1180 Vienna, Austria; stefan.vogel@boku.ac.at
[*] Correspondence: andreas.melcher@boku.ac.at; Tel.: +43-01-47654-93411

Received: 30 March 2020; Accepted: 19 May 2020; Published: 26 May 2020

Abstract: Water resources are among the fundamental resources that are the most threatened worldwide by various pressures. This study applied the Driver–Pressure–State–Impact–Response (DPSIR) framework as an innovative tool to better understand the dynamic interlinkages between the different sources of multiple stressors on aquatic ecosystems in Burkina Faso. The triangulation of evidences from interviews, literature reviews, and strategic simulations shows that several human impacts as well as climate change and its effects (such as the decrease of the water level, and the increase of the surface water temperature) are detrimental to fish productivity, abundance, and average size. Furthermore, the ongoing demographic and nutritional transition is driving cumulative pressures on water and fish resources. In this context, the development of aquaculture could offer alternative livelihoods and help fish stocks in natural ecosystems to recover, thereby reducing fishermen's vulnerability and easing overfishing pressures. Further, the empowerment of the actors and their participation to reinforce fisheries regulation are required to escape the current "regeneration trap" and to achieve a sustainable management of aquatic ecosystems in Burkina Faso.

Keywords: fisheries management; multiple stressors; stressors interaction; anthropogenic pressures; climate change; DPSIR; strategic simulations; interviews; river systems; Burkina Faso

---

1. Introduction

Although a tiny fraction, 0.01%, of the world's water and about 0.8% of Earth's surface, fresh water supports 6% of all species (100,000 out of 1.8 million) described for aquatic or terrestrial ecosystems [1]. This makes both inland water and its biodiversity an essential resource [1,2] for the economy, landscape, science, and education [1]. However, this extremely valuable resource is increasingly threatened on the planet because of global change [1–5], the human global footprint, and the number of humans, with their concomitant demand for this resource for food, health, and clean water [6–9]. Worldwide, rivers are among the most threatened ecosystems; multiple human pressures, such as pollution, water abstraction, river channelization, damming [10,11], and their complex interference with natural processes, are key to understanding the continuous degradation of such water resources [11].

As far as fisheries are concerned, global estimates reveal an increasing overexploitation; fish landings from inland waters have increased 400% since 1950, and many freshwater stocks are at risk of collapse [7,12]. In 2016, excluding aquaculture (aqc) production, Africa's inland capture production reached nearly 2.9 million tons, accounting for 25% of the global catches, and ranking the continent second worldwide after Asia (7.7 million tons produced) [13].

The issues related to water and fish have a special resonance for Burkina Faso (BF), a West African country located in the Sahel, a region which stretches from Senegal eastward to Sudan [14]. BF is a landlocked country bordered by six countries viz. Mali in the North and West, Niger in the East, and Ivory Coast, Ghana, Togo, and Benin in the South. The Burkinabe population exceeded 19 million inhabitants in 2017 with an average annual growth rate of 3.1% [15,16]. The population is highly reliant on agriculture as a means for subsistence but also as a source of income [17]. Indeed, the gross domestic product (GDP) is mainly based on the primary sector, including agriculture, livestock, forestry, and fisheries [18]. In 2018, the GDP per capita reached 715.12 USD [19]. In 2017, BF ranked 183rd out of 189 countries. Its very low human development index of 0.423 [20] is the expression of a widespread poverty that increases the vulnerability of the populations to climate change impacts, especially in rural areas, as more than 80% of the population live directly on subsistence agriculture [20,21].

With two-thirds of the country's climate being arid to semi-arid and only a few perennial waterbodies that persist throughout the year [22], the construction of permanent or temporary reservoirs has been perceived as a promising way to develop water resources to address water scarcity and improving food security in BF, notably in highly populated areas (e.g., around the Nakambe river) [22,23]. In 2018, nearly 1700 reservoirs were identified and listed [24,25], ranking BF as the country with the highest density of reservoirs in West Africa [26]. About 90% of these reservoirs do not exceed a storage capacity of 1 Mm^3 (million cubic meters) and are referred to as "small reservoirs" [27,28]. The water sources, especially groundwater and reservoirs, depend on rainfall [29]. Moreover, in a time of global climate warming, the integrity of these water sources is threatened by rising trends of increasing ambient air temperatures, which cause water loss due to evaporation, especially in the Sahelian zone [4].

Very few recent studies have been carried out on fish assemblages in BF [11,28,30,31], and the resulting lack of knowledge hinders efficient conservation measures and conceals the economic and ecological roles of fish [32]. The knowledge gap regarding multiple stressors in the freshwater ecosystem is a critical hindrance to efforts to establish sustainable inland fisheries. The interactions and cumulative effect of multiple stressors have been scarcely studied [2,7], while much research has been carried out about the individual effects of five key pressures on the freshwater ecosystem, including water extraction, habitat degradation (fragmentation and chemical pollution), over-exploitation, invasive species, and climate change [7,8,12,28,33]. Considering the increasing threat water resources and aquatic ecosystems are subject to, there is an urgent need for research [2] to fill knowledge gaps in freshwater ecosystems [2,3,34–36], especially in semi-arid and resource-poor countries, such as BF.

The overall goal of this paper was to understand the effects of the combination of multiple socio-ecological stressors on the integrity and performance of aquatic ecosystems in BF, i.e., the main tributary to the Volta river system, the Nakambe River or White Volta, and to analyze ways to improve the management of fisheries in the context of climate change. Furthermore, we explore how fish farming increases our capacity to address the twin challenges of aquatic ecosystems' degradation and food insecurity in BF.

2. Methods

2.1. Overarching Approach

A qualitative approach was applied to identify and explore new findings and to flexibly adapt how research was conducted along multiple parallel pathways [37]. Fishers and aquaculture (aqc) farmers as well as decision makers and staff members of state services at local, provincial, and national levels were involved in the study as informants but also, from a participatory perspective, as contributors to our common goal of defining future pathways toward a sustainable management of aquatic ecosystems.

We developed, assembled, and used multiple lines of evidence [38] across the Drivers–Pressure–State–Impact–Response (DPSIR) framework. Qualitative data from literature reviews, interviews, and strategic simulations were combined (Figure 1).

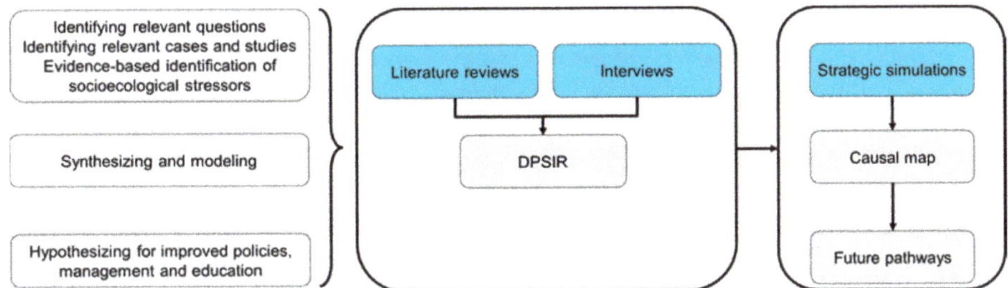

Figure 1. Multiple lines of evidence. This figure describes our approach of multiple lines identification of socio-ecological stressors in Burkina Faso's aquatic ecosystem. The Drivers–Pressure–State–Impact–Response (DPSIR) serves as the analytical framework shaping this research. The different research steps are displayed on the left side. The different methods or lines of evidence are highlighted in blue and the outcomes in grey. Literature reviews and interviews were used to elaborate a DPSIR model. This contributed as a background to the strategic simulations that yielded a causal map, and future pathways toward sustainable aquatic ecosystems and fisheries.

2.2. Literature Reviews

The aim of the literature reviews was to take stock of previous studies and gather evidence related, on the one hand, to the causal relations (drivers, pressure, state, impact, and responses) in BF water bodies and fisheries and the development of aqc, and, on the other hand, to identify the main adverse effects of climatic variabilities (climate change) on water and fish resources. Further, it helped in preparing relevant research questions and an interview guide.

A systematic review of published articles and grey literature was conducted using the Title-Abs-Key strings: ("Burkina Faso" OR "Niger" OR "Mali" OR "Benin" OR "Togo" OR "Ivory Coast" OR "Ghana" OR "West* Africa" OR "Sahel") AND ("climat* chang*" OR "global change" OR "global warming") AND (fish AND NOT marine); aquaculture AND ("Burkina Faso" OR "West Africa"). The search was carried out on Scopus for scholarly peer-reviewed literature and various databases for articles published in journals that are not indexed in Scopus as well as grey literature, such as Google Scholar© and Google©. A "snowball search" was used to include articles from reference lists of relevant publications and articles based on known literature and recommendations from colleagues. The same strategy was applied to search for literature in the French language, including published or grey literature. Once references in articles were observed to be particularly redundant, the snowball and grey literature searches were stopped.

The papers chosen to be reviewed focused on abiotic or biotic factors related to fish resources and the environment, and to the pressures arising from climate change impacts. This included the effects of anticipated climate variability on hydrological regimes in the region, and the resulting observed responses of fish. In addition, this included papers focusing on fishing communities' reaction to adapt to such changes and to reach sustainable practices. Papers regarding West Africa were excluded if (i) they dealt with coastal fish biodiversity and estuarine ecosystems (since BF is a landlocked country), and (ii) the climate of the study area diverged too much from Burkina Faso's climate. The exceptions were papers considering the need for research to better comprehend the evolution of climate change impacts in the region. The search yielded 190 documents (134 articles and 56 grey literature publications). Out of these, 65 articles were included (39 articles and 26 grey literature documents). Data and analysis from this literature search identified the components of the DPSIR and the abiotic and/or biotic variables recognized as being affected by climate change to complete the DPSIR causal map. The abiotic variables here referred to various climatic factors (e.g., precipitation variability, air temperature) and to physical and chemical characteristics of waterbodies (e.g., water temperature, dissolved oxygen content).

2.3. Interviews

We performed 27 interviews. Eighteen expert interviews focused on (i) pressures on reservoirs and streams, (ii) the current state of aqc, (iii) the potential for and impacts of aqc, (iv) the constraints for aqc development, and (v) the first steps to support the development of aqc. Meanwhile, nine complementary interviews with experts and stakeholders addressed interviewees' knowledge about climate-driven factors and the source(s) of their information, impacts on fish, policy recommendations, and sources of information regarding climate change.

The 18 expert interviews took place in BF from January to February 2018. Key persons from BF whose field of expertise is strongly related to fisheries and aqc were selected during a meeting with the team of the Austrian SUSFISH-Plus project (Sustainable Management of Water and Fish Resources in Burkina Faso) [39–41] as experts for the interviews. Other interviewees were selected based on the "snowball principle" [42], i.e., they were recommended by previous interviewees or key informants. Among the interviewees, eight were directors or employees at the national and regional level in departments related to fisheries and aqc. Six were academics: Professors, post-docs, PhD, and master's students from the University of Ouagadougou and the University of Nazi BONI in Bobo-Dioulasso, with backgrounds in natural or sociological sciences, both related to the topic of fisheries. Some ministerial and institutional officials were SUSFISH-Plus team members too. Four interviewees were aqc operators. Nine of those experts were SUSFISH-Plus team members from BF.

As far as the nine complementary interviews with experts and stakeholders are concerned, they were performed during the strategic simulations in Ouagadougou (see Section 2.4.) with aqc and fisheries' stakeholders and experts, including the president of the Fishermen Association and the president of the Women Processing Fish Association in Koubri (about 30 km South from Ouagadougou), the traditional canton chief of Koubri, the secretary-general of the Sport Fishermen Association in Ouagadougou, a professor, a research engineer, a member of the General Directorate of Fish Resources, a member of the Ministry of agriculture and hydro-agricultural development, and the program head of the International Union for Conservation of Nature (IUCN)-Burkina Faso. Six of the interviewees are working on a country-wide scale, and three are working in Koubri.

The results of the systematic literature reviews were qualitatively summarized and triangulated with the outcomes of the interviews (Figure 1) to cross-check and, where possible, enrich the results, one from another, and to thereby identify and describe the various components of the DPSIR framework: The potential drivers, pressures, current state of waterbodies, impacts, and responses. Regarding the construction of the DPSIR, the changes described in both the literature and interviews, or highlighted as being of special concern in one of these sources, were considered as relevant and are shown in the DPSIR causal map.

2.4. Reference Analytical Framework: DPSIR

The DPSIR framework was originally elaborated by the European Environment Agency (EEA) in 1999. It is a comprehensive systems-thinking framework that encompasses cause–effect relationships between interacting components of social, economic, and environmental systems [43]. It helps to simplify complex reality, to promote communication in an illustrative way, and thus assists to close the gap between science and decision-making [44]. The DPSIR has barely been used in Africa, thus it is an innovative tool for policy recommendation that can be of great interest for African governance [44,45]. The DPSIR has proven to be effective in organizing and communicating complex environmental information for policy formulation; therefore, it has been adopted as an analytical framework for environmental assessment [45,46] to examine land cover change [45], to identify major fisheries problems, adopt standardized indicators, and improve fisheries management [44,47–50]. The implementations of the DPSIR are variable. For instance, in Africa, Agyemang et al. [45] used two complementary assessment techniques: Geographical Information Systems (GIS) and remote sensing to assess the state of the environment (land cover change from 1990–2004) and participatory research methods (interviews, focus group discussions, participants' observations) to collect and triangulate participants' views in order to review the results of the first phase (cf. GIS and remote sensing), verify the results obtained, discuss the driving forces, pressures, and impacts of the changes, and reflect on their future responses. Instead, Gebremedhin et al. [44] describe the drivers, pressures, state, and impact based on the literature whilst responses were based on the literature, personal experience, and informal communication with fishermen, experts, and scientists [44]. In the European context, Knudsen et al. [48] also applied a modified DPSIR model, emphasizing drivers–pressure–state (DPS), to identify drivers for fishing pressure on the basis of ethnographic fieldwork and interviews in the coastal areas, Samsun fisheries on the Turkish Black Sea coast.

In the current research, the DPSIR was used as a tool to process the data from the literature and interviews, to elaborate the DPSIR model, and then to structure the presentation of the results. The factors stressed in both the literature and interviews or highlighted as being of special concern in one of the data sources are shown in the model diagram. The same method was used to identify all components of DPSIR.

The components of the DPSIR were defined as follows based on the guidance document for the analysis of pressures and impacts in the framework of the Common Implementation Strategy for the Water Framework Directive [51] and Chu [7]:

- *Drivers* are agents or processes that dominate the system's dynamics in such a way that they are unavoidable factors of change to ecosystems or human activities, including social, demographic, and economic development. As such, they profoundly influence nature and society.
- *Pressures* are direct results of the drivers. They can be human actions in response to the driver that affect aquatic ecosystems or the effects of the driver in the case of natural drivers (e.g., change in water chemistry due to the application of pesticides).
- *State* consists of quantitative or qualitative indicators that describe a component of the ecosystem of interest. In this study, it refers to the condition of water body (i.e., physical, chemical, and biological characteristics) resulting from both natural and anthropogenic factors.
- *Impacts* correspond to the effects of changes of the state on the ecosystem components, such as fish kill and human well-being, including economic prosperity, safety, and cultural well-being.
- *Responses* are the measures taken to improve the state of water bodies and to ensure the provision of ecosystem services. They can be also policies to prevent, mitigate, or adapt to the impacts triggered by the alterations of environmental states.

2.5. Strategic Simulations

The strategic simulations focused on the Nakambe (White Volta), one of the main river basins of BF, where the country and its residents face important development and environmental challenges.

The implementation of the simulations involved two steps in defining issues and their relations; (i) the current situation, and (ii) the future pathways; both are further described in this section. Prior to these steps, a list of important factors in the context of the basin and fish harvest were prepared based on an in-depth literature search, including the abovementioned methods (see Sections 2.2 and 2.3.) (Figure 1). Then, a systems diagram or "causal map" was created to represent the dependencies between the various factors.

Strategic simulations, known also as policy simulations or policy exercises, is an interactive participatory tool to engage stakeholders, researchers, and policy makers in the development of strategic insight. The participants of policy simulations work with real-world data and issues that resemble or are exact to their experience [52]. In the strategic simulations approach, participants explore real policy issues, using design elements known from serious games, such as game boards or cards, to structure communication [53] as well as to include feedback that participants receive based on their decisions. According to Solinska et al. [54], this "spurs the dynamics of the face-to-face simulations and adds the sense of realism, urgency and fun". As a result, even stakeholders without relevant academic backgrounds are able to engage in highly complex situations [55].

The stakeholders participating in the strategic simulations explore possible "pathways" embedded in existing external scenarios [56]. Past evaluations of similar processes indicate that the approach can be quite efficient in the development of strategic decisions [57]. It prepares stakeholders for a wide array of outcomes while considering existing strategies and plans rooted in real data and participants' own experiences. Strategic simulations have been successfully implemented to support stakeholder engagement and the data analysis process in various areas of research and strategy development, for example, flood risk on the Thames river [58], extreme sea-level rise due to climate change [59], social aspects of river-floodplain management [60], and international management of global climate change [61], but to the authors' knowledge, SUSFISH-Plus presented the first instance of such tool being used in a context of aquatic ecosystems in BF.

The strategic simulations (SSs) provided an opportunity for stakeholders to meet and define the issues related to sustainable fisheries using a map of BF as a visual representation of the spatial distribution of major resources and problem areas in aquatic socio-ecosystems. This exercise simulates a science–policy dialogue over a few hours that in real life might have taken months to years. As such, this exercise enhances learning by compressing space and time to speed learning and helps stakeholders to co-develop a better mutual cognitive understanding and a deeper discussion of key issues to develop recommendations relevant both to science and policy.

The SS exercise was implemented during two workshops. The first workshop took place in Vienna in September 2018 and aimed at identifying the important challenges and opportunities within a discussion that included an expanded scope of indicators during the exercise with the SUSFISH-Plus team. The second workshop took place in Ouagadougou in February 2019. It convened 42 stakeholders from BF to jointly discuss desirable futures and pathways with regard to fish management and water security. Twelve of the participants were fishery stakeholders and decision-makers regarding water and fisheries, including fishermen, fish sellers, fish farmers, the representative of rural municipality, representatives of the General Directorate of Fish (Ministry of Animal and Fish Resources), and the General Directorate of Agriculture and Hydraulic (Ministry of agriculture and hydro-agricultural development), a traditional chief. Other 12 participants were students. Additionally, SUSFISH-Plus members participated as facilitators (6 persons), scientific committee providing expert assistance to the participants (3 persons), and observers to record the key information of the simulations (9 persons).

2.5.1. Current Situation

This step consisted in reviewing the current situation in the Nakambe Basin and its major challenges in the context of fisheries. The process consisted in representing a simplified visual format of the Basin (Figure A1). Thus, a predefined set of materials, including Nakambe basin maps and cards representing the *"entities"* (i.e., land use, natural entities, fish, water, energy, transportation, safety, waste management, telecommunication, society-related facilities, and infrastructure), *"processes (activities)"* (i.e., natural process, social/technological activities), and *"indicators"* (e.g., food production) were provided to facilitate discussions. The materials, including thematic areas, were chosen to provide sufficient information without excessively narrowing the participants' scope of exploration and breadth of choices. Subsequently, participants were asked to define smaller areas within the basin that could be characterized by a different set of cards. Each area was assigned one of the four *"land use categories"* (viz. environment and agriculture, fish, industry, and risk zones/hot spot). Participants were free to move the cards around the map and add or remove them to depict the situation according to their expertise and to add new factors (i.e., economy, society, environment, water, food, energy, and fish) that were not included in the simulations.

After describing the current situation, including the challenges, risks, and opportunities on the spatial representation of the basin, the participants moved to the next step, the so-called "causal map". The causal map presented causal relations between indicators. All key indicators and correlations between them were identified by participants in the Vienna workshop. During the workshop in Burkina Faso, participants assessed the level of each indicator and moving through the causal loop, translated the level of indicator into its impact on next one. Stakeholders were encouraged to propose changes to the causal loop; however, no suggestions were made by the participants. The initial rating on a scale from 1–5 (with 1 being the lowest, 3 neutral, and 5 the highest value) was recoded as follows: *Low* (1–2), *medium* (3), and *high* (4–5). The interpretation of the levels depended on the indicator. Indeed, the 5th level of pollution can be interpreted as being very negative, whereas the same level of education can be seen as a very positive achievement.

2.5.2. Future Pathways

The future pathways exercise was designed to develop visions for the Nakambe basin and beyond as well as pathways leading to such visions. Three different groups worked on three different focuses, viz. (i) water and environment, (ii) food and fish, and (iii) economy and energy, respectively, to develop three visions together with their corresponding pathways, starting from clear ambitious but realistic visions of what can be achieved. To reach desired and holistic visions, the focus priorities were not supposed to eliminate other important concerns. Therefore, the same cards, as in the description of the current situation, were used. The pathways were represented also on the "Causal Map" to allow participants to compare various scenarios and their outcomes. This means they could add new entities or remove previous ones from the map (Figure A2). They could also assume whether these elements represent opportunities or risks, and whether they improve or worsen the current situation.

3. Results

3.1. DPSIR: Identification and Characterization of Socio-Ecological Stressors

In this sub-section, we present the main drivers, pressures, state, impacts, and responses identified based on two evidence lines, viz. literature reviews and the interviews. The components of the model are graphically represented in the following DPSIR causal map (Figure 2) and a brief description of each component is provided hereafter.

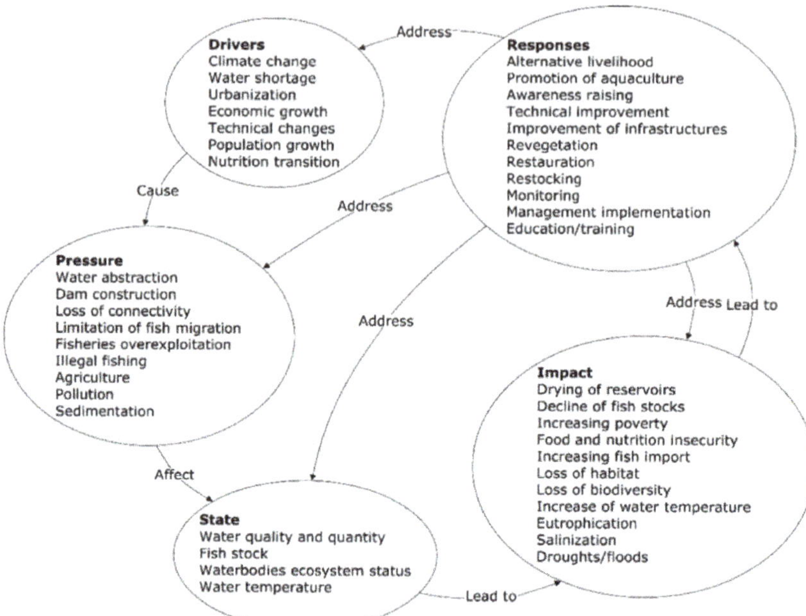

Figure 2. DPSIR relating to the BF aquatic ecosystem. The figure is a simplified representation of key factors in the interaction rather than the entire system. It is based on results from literature reviews and interviews. Source: Modified after Hundscheid [62].

3.1.1. Main Drivers

The drivers are essentially based on the literature as the interviews provided little material. Two types of transition occurred during the last century: Demographic and nutritional. Over these transitions, the current drivers of the multiple pressures on water and fish resources in BF have driven the overexploitation of water resources. Regarding the demographic transition, urbanization, technological changes in food processing, as well as economic growth led to the first demographic transition [63,64], which implies a significant increase in the population growth rate. In the current demographic transition stage, the rapid increasing population growth stems from decisive factors, such as an improved food and nutritional supply, better medical care, and a resulting reduced death rate, especially for children under the age of five, while the fertility rate remains constant [65]. Fertility levels in BF have declined only slightly over the past 50 years [66], thus the population has grown by 400% in the last 50 years. This transition has led to the increase in food and water demand, consequently, and a general overexploitation of natural resources in BF [67].

As far as the *nutritional transition* is concerned, it shifted the dietary patterns to a less carbohydrate-rich and more protein-rich nutrition [66,68]. The demand for fish as a protein source continues to rise because of the awareness of health benefits and lower costs, relative to meat, of fish [69].

Regarding environmental drivers, chronic and widespread water shortage is common to Sahelian countries [30]. Additionally, climate change exacerbates this issue through the decrease of rainfall, changes in the rain distribution patterns and surface flow rates, higher evapotranspiration driven by higher mean temperatures, and disturbance in the soil water balance, etc., resulting in silting of water bodies [25,62,70,71]. National experts expect average temperatures to rise by 0.8% by 2025 and by 1.7% by 2050. Meanwhile, the average annual rainfall is likely to decline by 3.4% by 2025 and by 7.3% by 2050 [62,70].

Human activities, including intensive agriculture, overgrazing, and uncontrolled clearing of land, also contribute to land erosion and worsen climate change impacts [67]. Above all, the overuse of water resources is a core driver of water shortage. The World Meteorological Organization revealed that the total annual demand of water in BF exceeds the available water resources by 10–22% and should continue to rise due to population growth [72].

3.1.2. Pressures

The four principal pressures consist of impacts of climate change (e.g., a decrease in the total annual precipitation, increase of mean annual air temperatures), dam construction, overfishing, and agriculture-related pressures. Regarding the impacts of climate change, both the literature survey and the interviews show that the abiotic indicators exhibiting the most variation due to climate change in the region include the significant decrease in total annual precipitation, the increase of mean annual air temperatures, and an increase in the inter-annual variation in precipitation [73–76]. The literature correlates well on this point with the observations made by different interviewees, such as fishermen, women fish processors in Koubri, a member of the General Directorate of Fish Resources, and the Traditional Canton Chief of Koubri. The intensification of evaporation, triggered by the increase in air temperatures, was often reported, as was the increase of drought frequency. The literature reviews as well as interviews show that abiotic indicators, such as water levels (i.e., volume of surface water available in aquatic ecosystems) is decreasing, as a result of climate change, and impact, in turn, fish (cf. biotic indicators) [74,77]. The major biotic indicators of indirect impacts on fish by climate change were decreases of fish abundance, fish productivity, primary production, and fish average size, and increases in the species extinction rate.

As shown in Table 1, a number of stressors associated with climate change (e.g., a decrease in flooding on floodplains (area and duration), decrease in water level and availability, decrease in dissolved oxygen content, and increase in surface water temperature individually) inhibit fish growth and, thereby, lead to a decrease in the average fish size and fish abundance. The decrease in the water level and availability and the increase of salinity results in a decrease in fish primary production as well as fish productivity. Secondly, both pollution and physical habitat modification (e.g., loss of connectivity, loss of habitats, habitat fragmentation) associated with dam construction affect the decrease in fish reproduction capacity while loss of connectivity blocks fish migration. Thirdly, a number of governance issues, e.g., lack of management implementation, illegal fishing, and ignorance of regulation, lead to decreases in the fish population, in fish biodiversity, and in fish average size. Finally, eutrophication, and a decrease in water quality and quantity, resulting from water abstraction for agriculture, cause a decrease in the fish population.

Table 1. Pressures, stressors, and their direct effects on the water ecosystem and fishes. The table was elaborated based on the results of the literature reviews realized from April 2017 to March 2018.

Pressure	Stressors on Freshwater Ecosystem	Direct Effects on Fishes
Climate change	Decrease in floods plains (area and duration) Decrease in water level and availability Decrease in dissolved oxygen content Increase of surface water temperature Eutrophication	Decrease in fish average size
	Floods plains (area and duration) decrease Decrease in water level and availability	Decrease in fish abundance
	Decrease in water level and availability Increase of salinity	Decrease in primary production
	Decrease in water level and availability Increase of salinity	Decrease in productivity
Dam construction (cf. habitat alteration and/or degradation through pollution and/or physical habitat change)	Pollution Physical habitat modification (e.g., loss of connectivity, loss of habitats, habitat fragmentation)	Decrease in reproduction capacity Block of fish migration
Overfishing	Lack of management implementation Illegal fishing Ignorance of regulation	Decrease in fish population Decrease in fish biodiversity Decrease in fish average size
Agriculture water abstraction	Eutrophication Decrease in water quality Decrease in water quantity	Decrease in fish population

Source: Authors' elaboration based on data from Hundscheid [62] and Paillaugue [78].

3.1.3. State

The main indicators of the current state of BF's water bodies consist of the waterbodies' ecosystem status, water quantity and quality, water temperature, and fish stocks. The ecosystem status of waterbodies declines as dam constructions lead to geomorphological changes to aquatic habitats, including the loss and fragmentation of habitats that serve as spawning grounds or habitats for juvenile fish stages of some species [79]. Consequently, species composition shifts such that both species richness [28] and fish stocks [44] decrease.

Similar changes to fish stocks result from the physiological stress caused by decreases of water quality and water quantity [79]. Decreasing water volume accelerates rising trends of water temperature and, hence, the deterioration of water quality. Water quality pressures occur especially in urban areas and agricultural areas [80]. Indeed, pollution through waste dumping in urban areas, pesticides and fertilizer input through agriculture, as well as eutrophication are detrimental to water quality.

The increase in plant and algae growth due to increasing water temperature and the entry of invasive fast-spreading plant species can lead to further eutrophication, oxygen reduction, and thus to impaired water quality. In addition, increasing water temperature can stimulate the growth of harmful algae blooms, which released toxins and may kill fish or trigger fish diseases [69]. An interviewee reported an additional effect of a "new bad aquatic grass" that was spreading very fast in the reservoir of Moussodougou with potential impacts on water quality and fishing activities.

Finally, overfishing is one of the main pressures leading to a decline in the total fish population, fish biodiversity, and average fish size. The latter is directly related to a reduction of the reproductive capacity [81], and thus resulted in a sharp decline of fish catches in the last years [70]. As a result, it is estimated that 56% of BF's fish species are currently threatened by the effects of human pressures [70,81].

3.1.4. Impact

The gap between rising evapotranspiration and decreasing rainfall creates a negative water balance that results in declining water levels [22]. This trend is exacerbated by rising rates of water abstraction. Farmers reported that during the dry season, some reservoirs no longer contain enough water to enable crop production [67], a trend that is corroborated by literature projecting that some reservoirs will likely dry out within the next decade [4].

Declines in the status waterbodies reflect critical losses of habitat that cause biodiversity losses and declines in fish stocks. Reduced fish stocks and the reduction of fish size have already led to a sharp reduction of quantity and quality of fish catches in many areas of BF [4,70]. Interviewees, from different administrative levels and sectors, national and regional institutions, as well as universities and fishermen, reported the continuous decline of fish stocks in reservoirs, lakes, and rivers. Additionally, the interviewees argue that declining fish stocks and reduced catches are triggering the increased usage of illegal fishing gears. Since fishing is perceived as "a fight against poverty", illegal fishing will continuously increase, failing any alternative policies to compensate for the loss of income and food or enhanced monitoring, policing, and enforcement.

The changes due to population growth and climate change are likely to have cascading impacts in BF. Socioeconomic impacts include increases in poverty and food and nutrition insecurity. These impacts are exacerbated by the expanding gap between rising population growth and fish demand and declining catches of indigenous fish. The latter leads to the last impact: Increasing fish imports. As a consequence, BF is highly dependent on fish imports. About 80% of the fish consumed in BF is imported [13], mainly from China, Taiwan, and neighboring countries like Ghana, Mali, and Ivory Coast.

3.1.5. Responses

Responses, i.e., actions intending to prevent, reduce, or adapt to pressures or environmental damage to improve BF's aquatic ecosystems (see Section 2.4.), include, first of all, education and training to raise the awareness and skill sets among resources' users. In particular, women's education can create new perspectives and independency, which, in combination with family planning can, in turn, lead to reduced birth rates [44,65,81] and induce the second demographic transition phase [68].

The implementation of international and national strategies for a sustainable fisheries management and aqc development is mandatory to meet the rising demand for protein and increase domestic fish production. Indeed, these strategies integrate international institutions and regulations, such as the Code of Conduct for Responsible Fisheries (CCRF), considered as a relevant guiding framework for implementing the principles of sustainable development in fisheries and aquaculture [13,82,83]. This will, in turn, reduce import dependency and boost the national economy. However, considering the overall lack of good fisheries management on regional and local levels [84], attention should rather be paid to improve the agency of fish managers through policy revision and improved implementation measures at sub-national levels rather than to the formulation of new national policies.

This lack of implementation can be traced back to the government's low deployment of resources for the fisheries sector at regional and local levels. Indeed, interviewees described the sector as "neglected". Our evidence, from interviews and literature, recommended that reversing this trend requires improved communication of the value of the sector to decision makers and implementation of multi-level governance that effectively functions at local and regional levels. In addition, improved collection of reference data would permit the long-term evaluation of policy performance [84]. Further, the collaboration of national institutions on different levels and international organizations is proposed as a way to develop and implement strategies for addressing climate change challenges.

Technical solutions to improve the sustainability of fisheries include the improvement of infrastructure and of the quantity and quality of critical habitats. A campaign to construct and validate fish ladders by management strategies could improve fish migration, which historically was much higher when most water bodies were rivers [11]. This improvement of connectivity is likely to raise fish diversity and abundance. Additionally, the revegetation and restoration of riparian buffer forests along the water bodies are suggested to improve water quality in agricultural areas and mitigate the siltation of reservoirs [4,80]. Furthermore, restrictions on fertilizer and pesticide usage should be introduced, and development towards more organic farming should be pursued.

Finally, to prevent further increases in poverty and malnutrition, experts recommended the development of alternative livelihoods to compensate for declines in incomes and food supplies due to declines in local fish catches. In this respect, both the literature and interviews confirm a high potential for aqc to provide alternative livelihoods that contribute to the recovery of fish stocks in reservoirs and to improve food security and livelihoods.

3.2. Strategic Simulations

The identification of socio-ecological stressors was done using another line of evidence, viz. strategic simulations. This exercise permitted participants (SUSFISH partners and fishery and aqc experts and stakeholders) to share their views of important stressors through an interactive tool. The results described the summary of stakeholders' work during the workshops.

3.2.1. Causal Mapping of the Current Situation

The key causal factors and their relations that influence the sustainability of fisheries in BF, which were revealed through the strategic simulations exercise, are mapped in Figure 3. This map shows that most of the indicators relating to wellbeing (including health security, food security, gender equality, and income) have a low level while high economic inequalities prevail. Further, stakeholder responses during the exercise reflected how wellbeing and development can be mutually co-dependent. For example, a low-income level hampers access to education, which, in turn, slows down the development in the area because it limits investments in specific technologies that depend on well-educated employees.

Development-related indicators also reflect new emerging anthropogenic pressures that can have a great impact on fish populations and fish harvests. For example, mining, especially the widespread open-pit mining in BF, increases pollution, both airborne and in surface runoff. Apart from biological pollution, which is low, heavy metal, chemical, and nutrient pollution from all sources have reached a "medium" level, as determined by expert opinion during the simulations exercise. Subsequently, high habitat degradation lowers species biodiversity, including fish, and hampers agricultural production by decreasing the organic content and bio-productivity of the soil, creating many hectares of barren land. Other development-related indicators (e.g., increasing population and urbanization) are also associated with pressures that are detrimental to the natural environment.

The Burkinabe economy has grown partly through the development of water infrastructures, such as man-made reservoirs, that secure water supplies for urban, industrial, and agricultural uses. However, these also might reduce fish habitat by creating additional fish migration barriers. Because this blocks the frequency of genetic exchange between fish populations, this could drastically lower the biodiversity of rivers and speed up the decline of some fish species. The level of water availability, which is medium, is connected to weather seasonality (the increasingly unpredictable durations of and shifts between dry and wet seasons) and the state of the existing water infrastructure. Further, it affects the size of the natural fish habitat and fish, which, combined with the diversity of species and their sensitivity to pollution, will impact the fish harvest potential. The latter can be further hindered by unregulated illegal practices, such as off-season fishing and illegal fishing equipment, both estimated to be high. Moreover, such illegal activities negatively impact the reproductive capacities of the fish and are inefficient in the long term. In summary, excessively intensive fish harvests, as a key source of food in BF, will continue to degrade the fish habitat and biodiversity, food security, and wellbeing of the population.

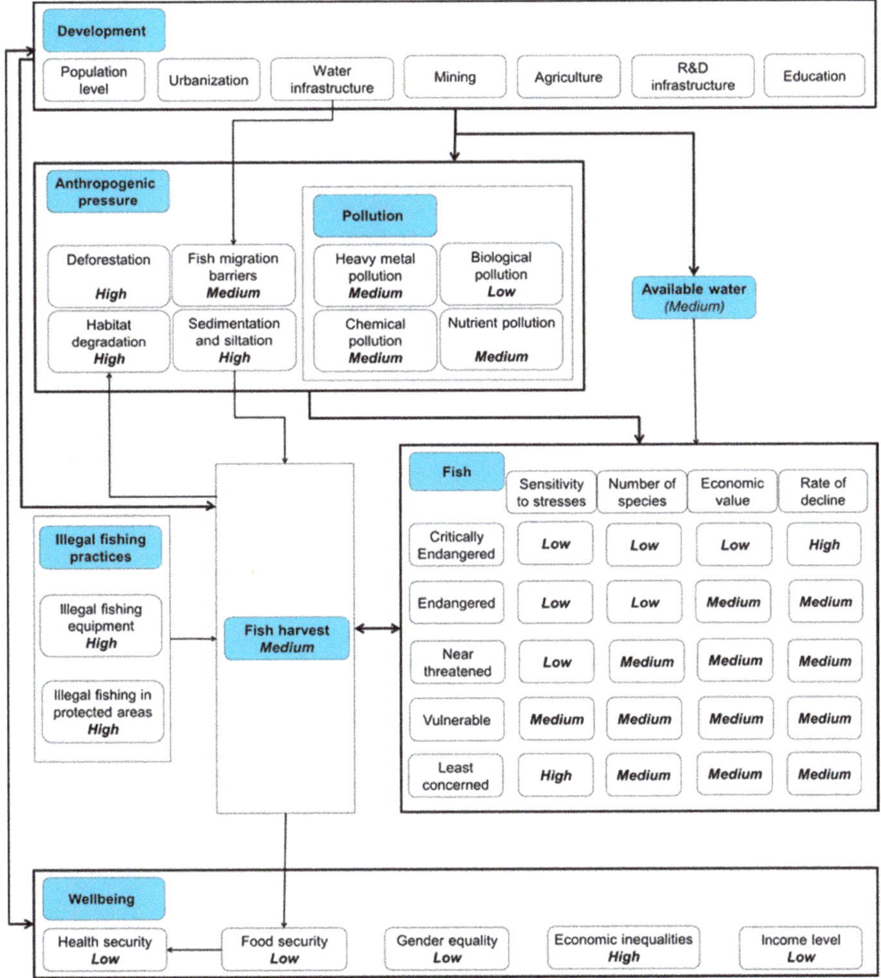

Figure 3. Causal map from the strategic simulations. The arrows show the interdependencies of the most important environmental and socio-economic factors of the Nakambe basin/BF system. The categories in italic and bold are the levels assigned by the participants to describe the current situation. The initial rating from 1 to 5 was recoded into *Low* (1–2), *Medium* (3), and *High* (4–5). R&D: Research and development.

3.2.2. Future Pathways for Sustainable Fisheries and Water Resources Management

The strategic simulations exercise provided three future narratives that describe visions and the pathways towards achieving these visions: (i) Vision-priority focus on the water and environment sector, (ii) vision-priority focus on the food and fish sectors, and (iii) vision-priority focus on the economy and energy sectors.

Vision-Priority Focus on Water and Environment Sector

According to the participants to the strategic simulations, in order to prosper, the Nakambe basin requires sufficient access to water and conservation of its natural ecosystems. The implementation of this vision demands, firstly, the collaboration between various agencies from different government levels and support from responsible ministries for water and environmental resources, including the Ministry of Animal and Fish Resources, the Ministry in charge of Water Resources, the Ministry in charge of Environment, and the Ministry in charge of Agriculture. Secondly, the improvement of water resources management requires investments in cleaning and protection of the waterbodies by shoreline reforestation and delimitation of the special buffer zones in the whole Nakambe basin. Thirdly, the educational system should be improved to raise awareness and develop dialogue between conflicting local stakeholders. Finally, subsidies for private owners and small businesses related to the fisheries could reduce the illegal practices and increase the contribution of the local enterprises to the local, and country's budget. Additionally, the reduction of the negative impact of illegal fishing in protected areas and off-season fishing on the biodiversity of fish may be realized by creating a more business-friendly environment and by popularizing fish farming. Indeed, aside from a very few protected aquatic ecosystems, almost all fisheries exhibit a small fraction of their productive potential.

Overfishing is not the result of one single cause. It emerges from many factors linked in multiple, reinforcing patterns of behavior, which have persisted partly because of a failure to properly establish and implement fisheries management policies [4] (see the sector marked actions in Figure 4). A constellation of reinforcing feedbacks has caught BF's fisheries in a regeneration trap" [85]. Indeed, fishermen realize the highest profit rates in the fish value chain. This attracts men from all sectors of rural society to try fishing for part of the year, no matter how unskilled they are. While competition and relatively low yields force most to rely on other sources of income, fishing continues to attract very high participation rates. Profitability can drive a feedback that reinforces increases in the number of fishermen and, hence, fishing rates and harvest income, which feeds back to increase profits. Unchecked, this reinforcing feedback loop will continue to drive all values higher and higher up to the point of fisheries collapse. Such a reinforcing feedback loop increases the number of fishermen and sustains fishing rates too high to permit fish populations to restore their natural reproductive capacity (Figure 4).

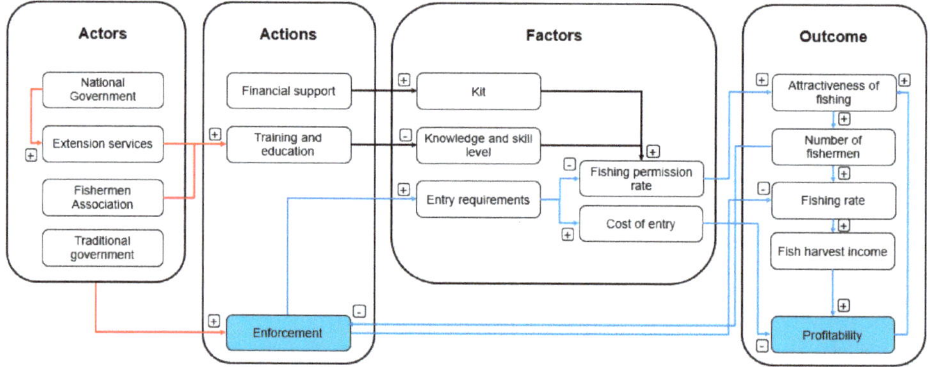

Figure 4. Governance failures leading to overfishing. The first variable to change is at the origin of the arrow, then the variable at the receiving end of the arrow will respond with a change. Regarding the "polarity", "+" shows that an increase or decrease in the first variable is matched by the same change in the receiving variable, "-" shows that the variables change in opposite directions. Actors: stakeholders for enforcement; actions: current measures; factors: elements that potentially could be provided by enforcement, training, and financial support to improve fishing governance; red arrows: interactions that are not established but could help make fisheries sustainable if instituted and established as standard parts of governance; blue arrows: feedback loops; outcome: potential results in terms of profitability; kit: fishermen' equipment. Source: modified after Sendzimir [85].

Vision-Priority Focus on Food and Fish Sectors

The second vision described by the participants to the strategic simulations consists of an integration of the investments in water and fish infrastructures to ensure food security and meet fish demand without complete depletion of the fish resources in the Nakambe basin. This would involve a series of transformations towards green energy sources so as to lower the negative impacts of water and air pollution resulting from fossil fuel energy by-products on both water and fish stock. The use of existing dams to produce hydro-energy will favor the productivity of the post-harvest facilities and crop harvesting. Further, investments in the creation of new dams and the development of irrigation, especially drip irrigation infrastructure, can contribute to develop the agricultural sector. Investments in fish stocking practices and new spawning grounds can help to sustain fish resources. Moreover, alternative energy sources (solar, wind) will contribute to minimize the negative impact of hydro-energy generation on the fish population. The improvement of the educational system is a core action in order to establish a skill base in BF society that can attract industries that exploit new ideas and innovate. Likewise, investments in the research and development will enhance the new possibilities that technology offers. Increased awareness of challenges by the general public and the policy sector, as well as the technical knowledge and soft skills (such as negotiating, leadership, and teamwork skills) are most important in implementing changes and transitioning towards a more green and efficient future.

Vision-Priority Focus on Economy and Energy Sectors

The third vision suggested by the participants to strategic simulations focused on the maintenance and the rehabilitation of existing water infrastructure to ensure better flow and quality of the water used for agriculture. For the participants to the strategic simulations, new dams should be equipped with fish passages to mitigate the negative effects of man-made barriers on fish reproduction, and biodiversity. To support farmers and the development of agriculture, a plan is needed to improve water delivery infrastructure. This would involve advancing drip irrigation technologies and reforming existing canal irrigation into more efficient systems.

Energy production can be addressed by diversifying the sources of energy to provide a secure and steady supply of energy to the widely distributed network of households and production facilities in BF. Further, the harmful impacts of hydroelectric dams on fish could be addressed by applying ideas from other countries, like the installation of nets (mesh, grating) for fish collection in the dams. Additionally, since climate change mandates the decarbonization of the economy, private owners and energy sector stakeholders could seize this challenge as an opportunity to shift from fossil fuel-based energy generation to cleaner solar energy. Since widespread open-pit mining fosters rural exodus and decreases the acreage of arable land, new employment policies could target youth and help adapt land newly reclaimed from mining to productive agricultural areas and invest in industrial mining. Finally, in this pathway, education is critical for the development of energy and economy sectors. Accordingly, curriculum reform should emphasize sustainable development and natural resources management as part of raising the general awareness about the current challenges of the basin. Finally, family planning will be promoted to stabilize the birth rate and the health of BF's residents.

Further Questions

The final visions for the Nakambe basin revealed a few tendencies among participants. Although we fostered a realistic approach among stakeholders, some of the results have a clearly aspirational element to them. Participants focused on the positive aspects of development and how to tackle the currently existing challenges, sometimes leaving out some of their negative effects that in reality would require countermeasures. A few topics were skimmed over or neglected by participants, for example, the budget. This was possibly due to a lack of time, sufficient data, or the variety of perspectives and goals represented among participants. Despite this, the visions presented above serve as an important source of knowledge about the region, stakeholders and decision makers in the region, their perception of the stressors, as well as their main aims and priorities for development.

4. Discussion

Multiple lines of evidence, e.g., literature reviews, interviews, and strategic simulations with stakeholders, were integrated to efficiently collect and combine data and research results of restricted availability. Indeed, the increasing recognition of the importance of epistemological pluralism [86] has deepened our appreciation of the value of local and traditional knowledge. This value is realized in part when research makes accessible to science and policy, knowledge and values that were established over decades of local experience [86–88] and constitute the ground on which certain societies judge their options. Further, it constitutes a source of detailed information about ecosystem processes and changes over long-time frames [86,89–91]. Thus, it can be used, especially, in a context where there are no documented data [87,92,93], or when difficulties in accessing accurate data are prevalent, especially in Africa [87,94].

The literature offered a dense and complete record of data and analyses of factors related to the sustainability of fisheries in BF, including data on the variations of climatic factors. The interviews confirmed the patterns already observed from the literature. Indeed, resources users' experiences and observations can help to confirm system interactions that relate directly to their livelihoods [95,96]. However, they may lack an objective understanding of the underlying socio-ecological system (SES) processes, e.g., processes linking nature and society [95,96] or gradual changes would sometimes escape their perception [86,97].

4.1. Stressors Interaction on Water and Fish Resources (Freshwater Ecosystem)

Increasingly extreme weather patterns associated with climate change increase water loss due to increased evapotranspiration and, overall, the length of droughts, thereby exacerbating water shortages in BF. Increasing and interrelated demands for water and food from society, enhanced by demographic transition (including population growth), aggravates this water shortage. The nutrition transition specifically increases the demand for proteins and thus the demand for fish [68]. The response to these demands has aggravated damages of aquatic ecosystems. Evidence from the literature, interviews, and strategic simulations highlights the roles of the construction of dams, the expansion of irrigated agriculture, and overfishing as core pressures on the BF aquatic ecosystem.

Our results are similar in several respects to the study of Gebremedhin et al. [44]. They identify population growth as well as economic growth as the main drivers, and the major pressures consist of agriculture, dam construction, urbanization, and fisheries. However, our study describes more broad drivers, including climate change. It is worth noting that our study was extended to Sub-Saharan Africa and put an emphasis on the relation between fish and climate change. Further, our approach combines literature reviews with interviews. Instead, Gebremedhin et al. [44] focused on the Lake Tana in Ethiopia. Moreover, they describe the drivers, pressures, state, and impact based on the literature whilst responses were based on the literature, personal experience, and informal communication with fishermen, experts, and scientists [44].

Knudsen et al. [48] also applied a modified DPSIR, emphasizing drivers–pressure–state (DPS) to identify drivers for fishing pressure on the basis of ethnographic fieldwork and interviews in the Samsun fisheries on the Turkish Black Sea coast. These drivers include (i) fish demand and consumption (high-value fish); (ii) tax exemption on fuel price; (iii) increasing availability and adoption of new technology (electronic equipment); (iv) availability of fishing infrastructure; (v) structural flexibility allowing fishermen to switch between different regions, gear, and income generation activities; and (vi) poverty. Although the context of this study (i.e., costal fisheries) is different from our study and that of Gebremedhin et al. [44] (i.e., inland fisheries), they all highlight techniques adoption, fishermen resources poverty and hence dependence, the economic value of fisheries, and their importance in nutrition as sources of fishing pressure [44,48].

The strategic simulations suggested that industrial mining could replace open-pit mining as a policy response to the pollution resulting from the latter's activity. However, even though this activity can result in environmental damage, the literature highlights its socio-economic importance, especially for local households in BF [98,99]. For instance, Pokorny et al. [98] argue that artisanal gold mining can generate job opportunities and cash income for local households. Bazillier and Victoire [99] show that a 1% increase in the gold price leads to a 0.12% increase in the consumption of households located close to artisanal mines. This supports a strong positive impact of artisanal mining on consumption. They further found a marginally significant amelioration in children's health associated with the income effect of artisanal mines that may outweigh any pollution effect. Both surveys suggest that industrial gold mining has failed to exert a positive influence on local households compared to artisanal mining even though it may contribute much more at the more macro level, to the state budget by generating significant proportions of urgently needed revenues and taxes [98,99].

Driven in part by climate change, declining water levels in aquatic ecosystems are impacting, in turn, fish abundance and average size. While increasingly variable climate patterns are key to understanding current and future potential threats to waterbodies and fish in BF, the trends cannot be generalized accurately to the whole country due to the very divergent climates from North to South. Indeed, the overall reliability of the climatic projection in West Africa and BF are challenged by the lack of observation data and monitoring in the region [12,100,101], and the lack of adequate tools to assess the impacts of climate change on water resources at the local level [102]. Most projections are not developed in ways that can usefully be scaled down and inform science and policy at basin scales in West Africa. They offer globalized analysis, using large-scale resolution for their climatic models, averaging several hundreds of kilometers, sometimes including the whole country in one pixel of resolution [103–105]. To address this issue, BF should strengthen its climate-related information system by adopting a systematic approach for coordinating the development, archiving, and use of such information. As for climate development data, it is important to modernize the network of the meteorological stations (e.g., automatic weather stations). It is also important to strengthen the National Meteorological Agency (ANAM) as well as the network of meteorological stations across the country. In addition to modernizing equipment and infrastructures, it is essential to strengthen the human capital by upgrading the competences and building the capacities of the human resources dealing with climate information at national, regional, and local levels. In this context, it is crucial to consolidate the collaboration with the World Meteorological Organization (WMO).

It is almost impossible to dissociate climate change impacts from direct anthropogenic pressures on waterbodies and fish. Several papers emphasized that all the observed changes in aquatic systems and fish resources are caused by the combination of climate change and direct anthropogenic pressures, such as land degradation (e.g., deforestation, soil pollution), rather than by climate change alone [4,73,81,106]. For instance, Darwall et al. [107] imputed the decline of the fish average size and the shrinking of their distribution to climate change, water abstraction, dams construction, and overfishing.

Our results highlighted overexploitation as a pressure, which corroborates previous studies in BF [28,41,108,109]. However, there is no unanimity in the scientific literature about the 'over-exploitation narrative' of freshwater fisheries [7,109,110]. Béné [109] and Béné and Russell [110] argue that, although the overexploitation in Lake Volta had already been suggested as a key factor more than 30 years ago and has been systematically brought forward in the past literature, the few time-series data available of fish landings over more than three decades do not substantiate this conclusion. Further, they highlighted that the discrepancies between official figures and the results of stock assessments may question the estimates, including the most rigorous ones. This is partly due to the lack of accurate data, including fish production figures [109,110], especially in BF. Chu et al. [7] also found that freshwater fisheries' overexploitation was not integrated as a pressure in the studies they reviewed.

Because of their structural complexity, the responses of animal and human communities may not be direct reactions to the rising trends of pressures and stressors. Despite the simple number of stressors in the ecosystems modified by humans, the interactions are complex with regard to the intensities and the temporal variation between them [111]. Secondly, tolerances of the fish population, derived from evolution or ecology, can modify biotic responses, such as biomass or species richness. In addition, biota can adapt to stressors and, likewise, biotic interactions can change at different stressors' levels [3,8,112]. Responses may vary within and between species, assemblages, and ecosystems; for example, headwaters compared to the middle of streams [113]. Thus, biotic interactions and responses may be not only linear but also non-linear or lagged [114].

Multiple lines of evidence enabled us to identify a diverse but not exhaustive list of pressures and stressors on BF aquatic ecosystems. Our results (Table 1) show that multiple stressors produce the same effect; therefore, it is likely that their combination implies interactions that amplify the effect of each other [7,8,115,116]. Evidence from marine environments already has shown that synergistic interaction between stressors, such as rising temperatures, biological invasions, and habitat destruction, exacerbates biodiversity loss and ecological degradation [3,8,117,118]. However, after analyzing three studies, Chu [7] found no consistent pattern; therefore, he highlighted that for additive, antagonistic, or synergistic interactions in multiple stressor assessments for freshwater fishes, interaction types depend chiefly on the species and life stage of interest, characteristics of the study ecosystem, indicators, and stressors (the same pressure can produce multiple stressors that affect state variables or indicators in different ways) [7]. We cannot on the basis of this study conclude whether the stressors in play are additive, synergetic, antagonistic, or an ecological surprise [7]. Therein lies a fundamental limitation of this research. Indeed, the study used a qualitative approach to understand stressor effects initially one by one, and then combined them [3,7]. This approach was motivated from a pragmatic point of view due to the difficulty to get and analyze accurate data specific to the species and the stressors.

Therefore, there is a need of further in-depth research using high-quality data with a wider scope (e.g., resolution in space and time) to test these hypotheses related to stressor effects and their interactions in the future. That could be done through experimental studies that compare controls to impacted ecosystems or replicate ecosystems exposed to similar stressors or gradients of stress [7]. Alternatively, meta-analyses could be performed [2,7,8,114].

The DPSIR has been subject to several criticisms, including (i) the risk of oversimplifying problems, (ii) the terminological unclarity, (iii) the lack of a methodological description to analyze disturbances [47,119,120], (iv) the lack of a consistent structure for systematically dealing with specific factors within the model components [121], and (v) the relative neglect of the more elusive deeper socio-cultural factors that underlie environmental theme problems [121]. Therefore, its implementation can be challenging for both policymakers and scientists. However, integrating the DPSIR with other conceptual frameworks [44,119,120] and more specific modelling tools [44,122] can effectively overcome these limitations. Further, notwithstanding the abovementioned limitations, the DPSIR has been proven to be a useful tool to organize and present complex information (i.e., the causality flow between human activities and nature) in a comprehensive way and to explicitly identify interventions or policy actions that can be taken at any level of the cause–effect relationships [43,44,121]. Consequently, it is a valuable approach that can yield knowledge, communication, and awareness for integrated and effective scientific, political, and public decision-making processes regarding sustainability or other key societal objectives [121]. However, as argued by Chu [7], the suggested responses are only useful to minimize, not eliminate, stressor effects. The complete mitigation of stressors requires the elimination of the corresponding drivers, which is often impossible [7].

4.2. From Stressors' Identification to Management Options

Considering the impacts of such stressors, management options should integrate measures to address population growth, the enforcement of international and national strategies for sustainable fisheries management, and the promotion of alternative livelihoods for poverty reduction and sustainable food security, such as sustainable agriculture and aqc.

The first policy response could consist of measures for family planning and improved women's education to reduce population growth, as proposed by the African Union [44,81,123]. At the national level, to tackle the problem, the State has set up several population policies, including the 1991 National Population Policy (PNP), which was revised in 2000, to reach a balance between population growth and socio-economic development [124]; and the National Acceleration Plan for Family Planning to increase the rate of modern contraceptive prevalence from 22.5% in 2015 to 32% in 2020 and accelerate the demographic transition [125]. Zan [124] shows that the actions aiming at reducing the population resulted in a decrease of the syncretic fertility index, with the highest decline observed in women with higher fertility, that is to say women with no education, with primary education, or living in rural areas. In addition, the syncretic fertility index is decreasing more in women who do not use modern contraceptive methods (62.9% against 37.10%). The author argues that this confirms a tendency to reduce fertility [124]. This observation suggests, he continues, the existence of traditional or natural methods to avoid pregnancy [124]. However, population and family planning policies have contributed much more to knowledge of contraceptive methods than to their use [124]. The adoption of family planning remains dependent on socio-economic factors, including the low level of education of women, hence the need for other development policies (e.g., education, urbanization, economic development). The literature shows that an increase in women's education can lead to a decline in fertility, even in the absence of family planning policies, because educated women are favorable to a lower fertility [124,126]. Given the direct negative correlation between the years of schooling of women and the fertility rate [127], an efficient starting point is the improvement of education and job opportunities for women [68]. These factors can contribute to the independence and free agency of women, e.g., their entry and participation, especially decision-making power, in the business sectors as well as governance [128]. For instance, women are already actively integrated in the fishing value chain, especially, in the area of fish processing and selling [4,28,41]. Therefore, strengthening this sector will also improve job opportunities and stimulate the participation of women in decision-making processes and administrative management of fisheries.

The second recommended policy response is the improvement of the implementation of strategies for sustainable fisheries. This requires the participation and empowerment of all actors [81]. Figure 4 presents four potential actors (viz. national government, traditional governance, extension services, fishermen's associations) who might contribute to the "protector" function either individually or in coordination with each other. Enforcement can be administered by agents of the traditional government (*Kotigi* or *Tengsoba*), by officers and/or members of fishermen's associations, and by agents of the national government or their regional proxies. Indeed, although trends show that traditional authorities' power is declining due to the spread of world religions, urbanization, and strengthening of republican governance [4,129,130], they may revive under current national campaigns to decentralize. These may offer local traditional authorities the opportunity to have access to the decision-making sphere and to be among the strong actors [130] that significantly influence actions on the ground. Under the premise that republican governance hampers and widely lacks successful local policy implementation, the population loses trust of national authorities, and thus traditional local chiefs are regaining power [131]. They can be key persons for the improvement of communication between the local population and republican governance. This, in turn, could strengthen the integration of local knowledge and participatory approaches in management strategies [132] if local traditional authorities could be included. As for the extension services provided by republican government, they function in many nations as experts to train farmers, fishermen, and other users of natural resources in the skills needed to sustainably use those resources. Such agents are almost wholly lacking in BF's fisheries sector.

Proper enforcement, as shown in Figure 4, might imply that police controls in aquatic areas when fishing occurs, limit the fishing rate as well as how fishing occurs, controlling the kind of equipment ("kit") used, and the manner of its use. It can also control whether and how fishing occurs by granting licenses to a select a number of applicants and checking whether their knowledge and skill level qualifies them to fish. Such enforcement would further constrain fishing by raising the cost of entry to the fishing profession, thereby lowering the profitability, and, hence, the attractiveness of fishing.

The factors (e.g., kit, skills, knowledge, and entry requirements) are elements by which enforcement agents might enable or deny access to fishing. Profitability is a determinant, since fishing provides cash income. For instance, decreasing profits due to increasing competition among fish processors is leading to a growing number of women to switch to other activities, such as gardening [110]. Profit can contribute to reduce the number of fishermen and, hence, fishing pressure. However, the implementation is challenging, because the state can barely muster the resources (personnel and equipment) to monitor and sanction illegal practices [41,130]. Previous studies in BF [41,130,133] show that organizing fishermen in associations and training them were successful instruments in implementing enforcement measures. As a result, fishermen associations respect such elements of governance control as limitations on the gear and mesh sizes, closed fishing seasons, and they collaborate in the monitoring, and add local initiatives for protecting aquatic resources. However, these studies further highlighted that the enforcement of fishing rights resulted in conflicting situations, because communities feel excluded from their resources for the benefit of individuals who have legal rights of use. This generated an increasing number of outsiders who trespassed on fishing grounds, especially in the case of concessions, which are granted as an "exclusive fishing right" [130,133]. However, as a policy open access also poses a risk for "tragedy of the commons" in fisheries [133]. Fishing is often only one part of the set of strategies of rural populations to diversify their economic activities in order to cope with risk [110,134]. Therefore, limiting access to fishing or reducing fishing profit can harm other activities and increase poverty levels among fishermen. Finding the balance between fishermen's needs and the preservation of aquatic resources is key.

In contrast to enforcement, the literature underlines incentives for current participants to leave the fisheries as a recommended policy to conserve aquatic resources [134]. However, the improvement of credit access would be conditional for the success of incentives. Indeed, the lack of formal credit facilities forces fishers to turn to informal credit providers, facing often an exploitive situation [41,110]. The State can create better conditions to increase opportunities, enable environment for investment, credit at low rate, simplify the legal framework, and provide training. These measures should raise the awareness among stakeholders about the positive effects of sustainable natural resources management and at the same time empower them in this area. Indeed, despite slight progress, the literature has reported a limited ecological awareness of local actors [41,108,135].

The third recommendation is the provision of alternative livelihoods, such as aqc and agriculture, to compensate reduced incomes, and to prevent further poverty and food insecurity. Béné and Russell [110] demonstrated in Volta Lake (Ghana) and in Bagré (BF) that the diversification of income sources is a determinant of fishermen's poverty status. Indeed, they argue that the fishermen depending on only fishing are among the poor group whilst the better off have diversified activities, including on- and off-farm activities. First of all, thousands of people in BF are dependent on fisheries [136] for income and subsistence. The secondarily generated income through fishing enables them to finance their medical supply, schooling for their children, and investments in agricultural activities [4]. Without alternative livelihoods that generate consistent income, people will increase their fishing activity and the use of illegal fishing techniques, regardless of fishing laws. Therein, the development of aqc has a high potential [69,70,137,138] and is regarded as "(. . .) the sole solution for boosting fish stocks in the Sahel region" [70]. However, direct and indirect interactions between aqc practice and the environment needs to be taken into account [139]. Besides the recovery of fish stocks by reducing the pressure of overfishing and proper active restocking measures [140,141], it can contribute to the provision of an alternative livelihood and protein supply. However, this approach should comply with the national regulation regarding the introduction of new species; the native fish population could be promoted rather than new alien invasive species. Indeed, the introduction of new species can either lead to an adaptation to ecological conditions; non-adaptation of the species, which results in its disappearance [4]; or other undesirable effects [108]. The failure of restocking can be attributed to fishermen's insufficient knowledge about fish ecology or even overfishing [108]. Hence, there is a need for capacity building for fishermen through training and better monitoring of the fisheries during the restocking process. In addition, the implementation of the aqc must be accompanied by an environmental impact assessment.

Secondly, since agriculture is the main activity of the majority of BF's rural population, including fishermen [41,108,109], the development of this sector can limit the rush over fish resources. However, considering the current pressure of agriculture on the aquatic ecosystem, the diffusion of more sustainable agricultural techniques is required.

5. Conclusions

Our study contributed to the identification of multiple stressors and their interactions that influence the sustainability of inland fisheries in BF's aquatic ecosystems. This was done by combining evidence from literature reviews, expert interviews, and strategic simulations with fisheries stakeholders. It allowed the construction of a causal map of the drivers, pressures, state, impacts, and responses using the DPSIR framework and a causal map of the current situation leading to future pathways about the water and fisheries management in BF. By providing simple causal maps, independent of vocabularies specific to separate disciplines, both models enable closer communication between researchers and decision-makers and open a debate about a holistic approach towards sustainable fisheries and aquaculture development to meet urgent needs to secure livelihood and food access for the population in BF. Such tools are useful in integrating inputs from stakeholders at all levels and in different sectors of the society. In this case, these models helped to better combine water- and fish-related ecosystem services' contributions to human well-being and the need to preserve these resources in the long term through a better knowledge (both indigenous and scientific) of their interaction and possible effective societal responses. This, in turn, highlights the importance of restoring the correct functionality of aquatic ecosystem services, especially in rivers belonging to very large catchments and on which many different geographical, socio-cultural, and political realities co-exist (e.g., Volta River flowing from BF into Ghana).

The water level was confirmed to currently decrease and to impact, in turn, fish abundance and average size, as a result of the current decrease of the total annual precipitation and the increase of annual precipitation variability. However, the uncertainties regarding the future precipitation patterns in the region (cf. Sahel/West Africa) may invalidate this hypothesis in the near future. Fish productivity, abundance, and average size were shown as the biotic indicators that are most recognized as being impacted nowadays by climate change in BF. The massive construction of dams and intensive agriculture have affected the status of aquatic ecosystems and, hence, of fish. With regard to overfishing, in addition to the lack of enforcement, profitability seems a determinant of overuse, since it drives a feedback loop, which reinforces the increase in the number of fishers and, consequently, fishing rates and incomes, which feedback to increase profits.

The gained understanding, from diverse perspectives, on the effects of multiple socio-ecological stressors on aquatic (water, fish) ecosystems across temporal and spatial scales in BF can be used to improve the effectiveness, efficacy, and sustainability of the management of these ecosystems in the context of climate change. Indeed, given the multiple biological, physical, and socio-economic stressors that affect the structure and functioning of aquatic ecosystems in BF, it is urgent to implement the management options and mitigation measures outlined in the present paper to ensure the long-term sustainability of fisheries and their vital contribution to food security and livelihoods in the country. The insights provided by the present paper on socio-ecological stressors affecting aquatic ecosystems and implications for sustainable management of fisheries in BF are valid for the other countries in West Africa, Sahel region, and Sub-Saharan Africa at large. Nevertheless, further research is needed on the interaction of multiple stressors and their impacts on the state of aquatic ecosystems in arid and semi-arid regions, to provide sound evidence to policy- and decision-makers.

Author Contributions: Conceptualization, V.-P.S., P.T., J.C.R., H.E.B., L.J.H., M.K., P.M. (Piotr Magnuszewski), J.P., J.S. and A.H.M.; Data curation, V.-P.S., J.C.R., L.J.H., M.K., P.M. (Piotr Magnuszewski), J.P. and A.H.M.; Formal analysis, V.-P.S., J.C.R., L.J.H., M.K., P.M. (Piotr Magnuszewski), J.P. and J.S.; Funding acquisition, A.H.M.; Investigation, V.-P.S., J.C.R., L.J.H., M.K., P.M. (Piotr Magnuszewski) and J.P.; Methodology, V.-P.S., P.T., J.C.R., H.E.B., L.J.H., M.K., P.M. (Piotr Magnuszewski), J.P., J.S. and A.H.M.; Project administration, P.T. and A.H.M.; Resources, V.-P.S., J.C.R., L.J.H., M.K., P.M. (Piotr Magnuszewski), J.P. and J.S.; Software, V.-P.S., J.C.R., L.J.H., M.K., P.M. (Piotr Magnuszewski), J.P. and J.S.; Supervision, P.T., H.E.B., G.S., S.V. and A.H.M.; Validation, V.-P.S., J.C.R., L.J.H., M.K., P.M. (Piotr Magnuszewski), J.P. and A.H.M.; Visualization, V.-P.S., L.J.H., M.K., P.M. (Piotr Magnuszewski), J.P. and J.S.; Writing—original draft, V.-P.S., J.C.R., L.J.H., M.K., P.M. (Piotr Magnuszewski), J.P. and J.S.; Writing—review and editing, V.-P.S., P.T., H.E.B., M.K., P.M. (Piotr Magnuszewski), P.M. (Paul Meulenbroek), J.S., G.S., S.V. and A.H.M. All authors have read and agreed to the published version of the manuscript.

Funding: This research project SUSFISH-plus (project166) was funded by APPEAR, the Austrian Partnership Programme in Higher Education and Research for Development, a programme funded by the Austrian Development Cooperation (ADC) and implemented by the Austrian Agency for International Cooperation in Education and Research (OeAD), see also https://appear.at/en/. Grant number OEZA 0894-00/2014 and The APC was funded by SUSFISH-plus (project166).

Acknowledgments: We thank all interviewees for sharing their time, knowledge and opinions with us. We are also grateful to all members of the project Sustainable Management of Water and Fish Resources in Burkina Faso (SUSFISH-plus), in which this research is embedded, for their team work and spirit. SUSFISH-plus aims at establishing sustainable socio-economic fisheries and water management through the improvement of higher education and governance in Burkina Faso [39–41].

Conflicts of Interest: The authors declare no conflict of interest. The funders had no role in the design of the study; in the collection, analyses, or interpretation of data; in the writing of the manuscript, or in the decision to publish the results.

Appendix A

Figure A1. Participants describing the current situation of Nakambe Basin, during the strategic simulations in Ouagadougou.

Figure A2. Representation by the participants to the strategic simulations of the future pathways regarding the vision-priority focus on economy and energy sectors in Nakambe Basin.

References

1. Dudgeon, D.; Arthington, A.H.; Gessner, M.O.; Kawabata, Z.I.; Knowler, D.J.; Lévêque, C.; Naiman, R.J.; Prieur-Richard, A.H.; Soto, D.; Stiassny, M.L.J.; et al. Freshwater biodiversity: Importance, threats, status and conservation challenges. *Biol. Rev. Camb. Philos. Soc.* **2006**, *81*, 163–182. [CrossRef] [PubMed]
2. Ormerod, S.J.; Dobson, M.; Hildrew, A.G.; Townsend, C.R. Multiple stressors in freshwater ecosystems. *Freshw. Biol.* **2010**, *55*, 1–4. [CrossRef]
3. Jackson, M.C.; Loewen, C.J.G.; Vinebrooke, R.D.; Chimimba, C.T. Net effects of multiple stressors in freshwater ecosystems: A meta-analysis. *Glob. Chang. Biol.* **2016**, *22*, 180–189. [CrossRef]
4. Ouedraodogo, R. Fish and Fisheries Prospective in Arid Inland Waters of Burkina Faso, West Africa. Ph.D. Thesis, University of Natural Resources and Life Sciences (BOKU), Vienna, Austria, 2010.
5. Malqmqvist, B.; Rundle, S. Threats to the running water ecosystems of the world. *Environ. Conserv.* **2002**, *29*, 134–153. [CrossRef]
6. Sanderson, E.W.; Jaiteh, M.; Levy, M.A.; Redford, K.H.; Wannebo, A.V.; Woolmer, G. The human footprint and the last of the wild. *Bioscience* **2002**, *52*, 891–904. [CrossRef]
7. Chu, C.; Barker, J.; Gutowsky, L.; de Kerckhove, D. *A Conceptual Management Framework for Multiple Stressor Interactions in Freshwater Lakes and Rivers*; Ontario Ministry of Natural Resources and Forestry: Peterborough, ON, Canada, 2018.
8. Crain, C.M.; Kroeker, K.; Halpern, B.S. Interactive and cumulative effects of multiple human stressors in marine systems. *Ecol. Lett.* **2008**, *11*, 1304–1315. [CrossRef]
9. Halpern, B.S.; Selkoe, K.A.; Micheli, F.; Kappel, C.V. Evaluating and ranking the vulnerability of global marine ecosystems to anthropogenic threats. *Conserv. Biol.* **2007**, *21*, 1301–1315. [CrossRef]
10. Schmutz, S.; Sendzimir, J. Science for governing towards a sustainable future. In *Riverine Ecosystem Management*; Schmutz, S., Sendzimir, J., Eds.; Springer: Cham, Switzerland, 2018; ISBN 9783319732497.
11. Melcher, A.H.; Ouedraogo, R.; Schmutz, S. Spatial and seasonal fish community patterns in impacted and protected semi-arid rivers of Burkina Faso. *Ecol. Eng.* **2012**, *48*, 117–129. [CrossRef]
12. Allan, J.D.; Abell, R.; Hogan, Z.E.B.; Revenga, C.; Taylor, B.W.; Welcomme, R.L.; Winemiller, K. Overfishing of Inland Waters. *Bioscience* **2005**, *55*, 1041–1051. [CrossRef]
13. FAO. *The State of World Fisheries and Aquaculture 2018—Meeting the Sustainable Development Goals*; FAO: Rome, Italy, 2018; ISBN 9789251305621.

14. Encyclopaedia Britannica Sahel. Region, Africa. Available online: https://www.britannica.com/place/Sahel (accessed on 22 April 2019).

15. Climate Investment Funds Burkina Faso. Available online: https://www.climateinvestmentfunds.org/country/burkina-faso (accessed on 2 October 2018).

16. World Bank Burkina Faso: Overview. Available online: http://www.worldbank.org/en/country/burkinafaso/overview#1 (accessed on 2 October 2018).

17. Dresch, J.; Echenberg, M.; Guiguemde, P.H.; Deschamps, H.J. Burkina Faso. Available online: https://www.britannica.com/place/Burkina-Faso#ref54885 (accessed on 2 October 2018).

18. SP/CONEDD. *Evaluation de la Vulnérabilité et des Capacités D'adaptation aux Changements Climatiques du Burkina Faso*; SP/CONEDD: Ouagadougou, Burkina Faso, 2006.

19. World Bank Burkina Faso. Available online: https://data.worldbank.org/indicator/NY.GDP.PCAP.CD?locations=BF&view=chart (accessed on 30 January 2020).

20. UNDP Human Development Indices and Indicators: 2018 Statistical Update. Briefing Note for Countries on the 2018 Statistical Update: Burkina Faso. Available online: http://www.hdr.undp.org/sites/all/themes/hdr_theme/country-notes/BFA.pdf (accessed on 9 October 2018).

21. African Development Bank Country Strategy Paper (CSP) 2017–2021: Burkina Faso. Available online: https://www.afdb.org/en/documents/document/burkina-faso-country-strategy-paper-2017-2021-98046/ (accessed on 27 November 2018).

22. Fowe, T.; Karambiri, H.; Paturel, J.E.; Poussin, J.C.; Cecchi, P. Water balance of small reservoirs in the Volta basin: A case study of boura reservoir in Burkina Faso. *Agric. Water Manag.* **2015**, *152*, 99–109. [CrossRef]

23. Boelee, E.; Cecchi, P.; Koné, A. *Health Impacts of Small Reservoirs in Burkina Faso*; IWMI Working Paper; International Water Management Institute: Colombo, Sri Lanka, 2009; ISBN 978-92-9090-717-6.

24. Venot, J.-P.; Cecchi, P. Valeurs d'usage ou performances techniques: Comment apprécier le rôle des petits barrages en Afrique subsaharienne? *Cah. Agric.* **2011**, *20*, 112–117.

25. Jaime, C.R. GIS Based Analysis of the Environmental and Societal Importance of the Reservoirs in Burkina Faso. Master's Thesis, University of Natural Resources and Life Sciences (BOKU), Vienna, Austria, 2019.

26. Cecchi, P.; Meunier-Nikiema, A.; Moiroux, N.; Sanou, B. Towards an atlas of lakes and reservoirs in Burkina Faso. In *Small Reservoirs Toolkit*; Andreini, M., Schuetz, T., Harrington, L., Eds.; IWMI: Battaramulla, Sri Lanka, 2009; p. 23.

27. Lemoalle, J.; Condappa, D. *Water Atlas of the Volta Basin-Atlas De L'Eau Dans Le Bassin De La Volta*; Challenge Program on Water and Food and Institut de Recherche pour le Développement: Colombo, Sri Lanka; Marseille, France, 2009; ISBN 9789295090002.

28. Meulenbroek, P.; Stranzl, S.; Oueda, A.; Sendzimir, J.; Mano, K.; Kabore, I.; Ouedraogo, R.; Melcher, A. Fish communities, habitat use, and human pressures in the Upper Volta basin, Burkina Faso, West Africa. *Sustainability* **2019**, *11*, 5444. [CrossRef]

29. Ministère de l'Environnement et du Cadre de Vie. *Burkina Faso National Climate Change Adaptation Plan (NAP)*; Ministère de l'Environnement et du Cadre de Vie: Ouagadougou, Burkina Faso, 2007; p. 152.

30. Baijot, E.; Moreau, J.; Bouda, S. *Aspects Hydrobiologiques Et Piscicoles Des Retenues D'Eau En Zone Soudano-Sahélienne. Le Cas Du Burkina Faso*; Centre Technique de Coopération Agricole et Rurale (CTA): Brussels, Belgium, 1994; ISBN 9290811242.

31. Paugy, D.; Fermon, Y.; Abban, K.E.; Diop, M.E.; Traoré, K. Onchocerciasis control programme in West Africa: A 20-year monitoring of fish assemblages. *Aquat. Living Resour.* **1999**, *12*, 363–378. [CrossRef]

32. Mano, K. Fish Assemblages and Aquatic Ecological Integrity in Burkina Faso. Ph.D. Thesis, University of Natural Resources and Life Sciences (BOKU), Vienna, Austria, 2016.

33. Myers, B.J.E.; Lynch, A.J.; Bunnell, D.B.; Chu, C.; Falke, J.A.; Kovach, R.P.; Krabbenhoft, T.J.; Kwak, T.J.; Paukert, C.P. Global synthesis of the documented and projected effects of climate change on inland fishes. *Rev. Fish Biol. Fish.* **2017**, *27*, 339–361. [CrossRef]

34. Root, T.L.; Price, J.T.; Hall, K.R.; Schneider, S.H. Fingerprints of global warming on wild animals and plants. *Nature* **2003**, *421*, 57–60. [CrossRef] [PubMed]

35. Staudt, A.; Leidner, A.K.; Howard, J.; Brauman, K.A.; Dukes, J.S.; Hansen, L.J.; Paukert, C.; Sabo, J.; Solórzano, L.A. The added complications of climate change: Understanding and managing biodiversity and ecosystems. *Front. Ecol. Environ.* **2013**, *11*, 494–501. [CrossRef]

36. Hering, D.; Carvalho, L.; Argillier, C.; Beklioglu, M.; Borja, A.; Cardoso, A.C.; Duel, H.; Ferreira, T.; Globevnik, L.; Hanganu, J.; et al. Managing aquatic ecosystems and water resources under multiple stress—An introduction to the MARS project. *Sci. Total Environ.* **2015**, *503*, 10–21. [CrossRef]

37. Lamnek, S.; Krell, C. *Qualitative Sozialforschung*; Beltz Verlag: Weinheim, Basel, 2005; ISBN 9783621283625.

38. Kraft, K.H.; Brown, C.H.; Nabhan, G.P.; Luedeling, E.; Ruiz, J.D.J.L.; d'Eeckenbrugge, G.C.; Hijmans, R.J.; Gepts, P. Multiple lines of evidence for the origin of domesticated chili pepper, *Capsicum annuum*, in Mexico. *Proc. Natl. Acad. Sci. USA* **2014**, *111*, 6165–6170. [CrossRef]

39. APPEAR Sustainable Management of Water and Fish Resources in Burkina Faso. Available online: https://appear.at/en/projects/current-projects/project-websites/project166-susfish-plus/ (accessed on 10 February 2020).

40. Susfish Project Overview. Available online: http://susfish.boku.ac.at/contao/index.php/beispiel-startseite.html (accessed on 10 February 2020).

41. Slezak, G.; Sendzimir, J.; Ouedraogo, R.; Meulenbroek, P.; Savadogo, M.; Kabore, C.; Oueda, A.; Toe, P.; Zerbo, H.; Melcher, A. Fishing for food and food for fish—Negotiating long-term, sustainable food and water resources in a transdisciplinary research project in Burkina Faso. In *Towards Shared Research: Participatory and Integrative Approaches in Researching African Environments*; Zingerli, C., Haller, T., Eds.; Transcript Verlag: Bielefeld, Germany, 2020; pp. 125–165. ISBN 9783839451502.

42. Patton, M.Q. *How to Use Qualitative Methods in Evaluation*; Sage: London, UK, 1987; ISBN 0803931298.

43. Bradley, P.; Yee, S. *Using the DPSIR Framework to Develop a Conceptual Model: Technical Support Document*; US Environmental Protection Agency, Office of Research and Development, Atlantic Ecology Division: Narragansett, RI, USA, 2015.

44. Gebremedhin, S.; Getahun, A.; Anteneh, W.; Bruneel, S.; Goethals, P. A Drivers-Pressure-State-Impact-Responses framework to support the sustainability of fish and fisheries in Lake Tana, Ethiopia. *Sustainability* **2018**, *10*, 2957. [CrossRef]

45. Agyemang, I.; McDonald, A.; Carver, S. Application of the DPSIR framework to environmental degradation assessment in northern Ghana. *Nat. Resour. Forum* **2007**, *31*, 212–225. [CrossRef]

46. Niemeijer, D.; de Groot, R.S. Framing environmental indicators: Moving from causal chains to causal networks. *Environ. Dev. Sustain.* **2008**, *10*, 89–106. [CrossRef]

47. Martins, J.H.; Camanho, A.S.; Gaspar, M.B. A review of the application of driving forces-Pressure-State-Impact-Response framework to fisheries management. *Ocean. Coast. Manag.* **2012**, *69*, 273–281. [CrossRef]

48. Knudsen, S.; Zengin, M.; Koçak, M.H. Identifying drivers for fishing pressure. A multidisciplinary study of trawl and sea snail fisheries in Samsun, Black Sea coast of Turkey. *Ocean. Coast. Manag.* **2010**, *53*, 252–269. [CrossRef]

49. Mangi, S.C.; Roberts, C.M.; Rodwell, L.D. Reef fisheries management in Kenya: Preliminary approach using the driver-pressure-state-impacts-response (DPSIR) scheme of indicators. *Ocean. Coast. Manag.* **2007**, *50*, 463–480. [CrossRef]

50. Kelble, C.R.; Loomis, D.K.; Lovelace, S.; Nuttle, W.K.; Ortner, P.B.; Fletcher, P.; Cook, G.S.; Lorenz, J.J.; Boyer, J.N. The EBM-DPSER conceptual model: Integrating ecosystem services into the DPSIR framework. *PLoS ONE* **2013**, *8*. [CrossRef] [PubMed]

51. European Commission Analysis of Pressures and Impacts. In *Guidance Document No 3*; OPOCE: Luxembourg, 2003.

52. Harvey, S.; Liddell, A.; McMahon, L. *Windmill 2009*; O'Neill, K., Ed.; Dutch University Press: London, UK, 2009; ISBN 978 1 85717 588 2.

53. Duke, R.D.; Geurts, J.L.A. *Policy Games for Strategic Management*; Rozenberg Publishers: Amsterdam, The Netherlands, 2004; ISBN 9036193419.

54. Solinska-Nowak, A.; Magnuszewski, P.; Curl, M.; French, A.; Keating, A.; Mochizuki, J.; Liu, W.; Mechler, R.; Kulakowska, M.; Jarzabek, L. An overview of serious games for disaster risk management—Prospects and limitations for informing actions to arrest increasing risk. *Int. J. Disaster Risk Reduct.* **2018**, *31*, 1013–1029. [CrossRef]

55. Wada, Y.; Vinca, A.; Parkinson, S.; Willaarts, B.A.; Magnuszewski, P.; Mochizuki, J.; Mayor, B.; Wang, Y.; Burek, P.; Byers, E.; et al. Co-designing Indus water-energy-land futures. *One Earth* **2019**, *1*, 185–194. [CrossRef]

56. Van Notten, P. Scenario development: A typology of approaches. In *Think Scenarios. Rethink Education*; OECD, Ed.; OECD: Paris, France, 2006; pp. 69–92. ISBN 9789264023642.

57. Volkery, A.; Ribeiro, T. Scenario planning in public policy: Understanding use, impacts and the role of institutional context factors. *Technol. Forecast. Soc. Chang.* **2009**, *76*, 1198–1207. [CrossRef]

58. Lonsdale, K.G.; Downing, T.E.; Nicholls, R.J.; Parker, D.; Vafeidis, A.T.; Dawson, R.; Hall, J. Plausible responses to the threat of rapid sea-level rise in the Thames Estuary. *Clim. Chang.* **2008**, *91*, 145–169. [CrossRef]

59. Toth, F.L.; Hizsnyik, E. Managing the inconceivable: Participatory assessments of impacts and responses to extreme climate change. *Clim. Chang.* **2008**, *91*, 81–101. [CrossRef]

60. Stefanska, J.; Magnuszewski, P.; Sendzimir, J.; Romaniuk, P.; Taillieu, T.; Dubel, A.; Flachner, Z.; Balogh, P. A gaming exercise to explore problem-solving versus relational activities for river floodplain management. *Environ. Policy Gov.* **2011**, *21*, 454–471. [CrossRef]

61. Parson, E.A. *A Global Climate-Change Policy Exercise: Results of a Test Run, July 27–29, 1995*; IIASA Working Paper, WP-96-090; International Institute for Applied Systems Analysis: Laxenburg, Austria, 1996.

62. Hundscheid, L.J. Aquaculture in Burkina Faso—A Niche in Transition. Master's Thesis, University of Natural Resources and Life Sciences (BOKU), Vienna, Austria, 2019.

63. Thompson, W.S. Population. *Am. J. Sociol.* **1929**, *34*, 959–975. [CrossRef]

64. Kirk, D. Demographic transition theory. *Popul. Stud.* **1996**, *50*, 361–387. [CrossRef] [PubMed]

65. Mendez, M.A.; Popkin, B.M. Globalization, urbanization and nutritional change in the developing world. *Electron. J. Agric. Dev. Econ.* **2004**, *1*, 220–241.

66. World Bank. *The World Bank Group A to Z 2016*; World Bank: Washington, DC, USA, 2016; ISBN 978-1-4648-0655-1.

67. Sawadogo, J.M. Coping With Less Rain in Burkina Faso. Available online: https://www.un.org/africarenewal/magazine/july-2007/coping-less-rain-burkina-faso (accessed on 1 April 2019).

68. Popkin, B. What is The Nutrition Transition. Available online: http://www.cpc.unc.edu/projects/nutrans/whatis (accessed on 20 June 2018).

69. Satia, B.P. Regional review on status and trends in aquaculture development in Sub-Saharan Africa—2010. In *FAO Fisheries and Aquaculture Circular No. 1061/4*; FAO, Ed.; FAO: Rome, Italy, 2017; p. 191.

70. Amoussou, O.T.; Youssao, I.A.K.; Toguyéni, A. Improving aquaculture production in the Kou valley, Burkina Faso. In *Agricultural Innovations for Sustainable Development: Contributions from the 3rd Africa-Wide Women and Young Professionals in Science Competitions*; Francis, J.A., Ed.; Technical Centre for Agricultural and Rural Cooperation (CTA): Wageningen, The Netherlands; Forum for Agricultural Research in Africa (FARA): Accra, Ghana, 2014; Volume 4, pp. 187–194.

71. Awotwi, A.; Kumi, M.; Jansson, P.E.; Yeboah, F.; Nti, I.K. Predicting hydrological response to climate change in the White Volta catchment, West Africa. *J. Earth Sci. Clim. Chang.* **2015**, *6*, 1–7.

72. Savadogo, A.-S. Water Resource Management in Burkina Faso. Available online: https://www.wateraid.org/search/wasearch (accessed on 3 March 2020).

73. Diarra, D.Z. *Impacts Des Changements Climatiques En Afrique De l'Ouest*; Direction Nationale de la Météorologie: Bamako, Mali, 2009.

74. Stücklin, S.B.; Frei, K. Between water spirits and market forces: Institutional changes in the Niger inland delta fisheries among the Somono and bozo fishermen of Wandiaka and Daga-Womina (Mali). In *Disputing the Floodplains: Institutional Change and the Politics of Resource Management in African Wetlands*; Haller, T., Ed.; Brill: Leiden, The Netherland, 2010; pp. 77–119.

75. Barrat, J.-M. *Changements Climatiques En Afrique De l'Ouest Et Conséquences Sur Les Eaux Souterraines. Rapport OSS-GICRESAIT no 13*; Observatoire du Sahara et du Sahel (OSS): Tunis, Tunisia, 2012.

76. Yang, T.; Cui, T.; Xu, C.-Y.; Ciais, P.; Shi, P. Development of a new IHA method for impact assessment of climate change on flow regime. *Glob. Planet. Chang.* **2017**, *156*, 68–79. [CrossRef]

77. Touré, A. Changement climatique. In *Sédentarisation Et Adaptation Des Pêcheurs Du Delta Central Du Niger: Cas Des Communes De Mopti Et Konna*; Université de Bamako: Bamako, Mali, 2006.

78. Paillaugue, J. Climate Change Impacts on Fish Resources in Burkina Faso. Master's Thesis, University of Natural Resources and Life Sciences (BOKU), Vienna, Austria, 2019.

79. Schinegger, R.; Trautwein, C.; Melcher, A.; Schmutz, S. Multiple human pressures and their spatial patterns in European running waters. *Water Environ. J.* **2012**, *26*, 261–273. [CrossRef]

80. Stranzl, S. Quantification of Human Impacts on Fish Assemblages in the Upper Volta Catchment, Burkina Faso. Master's Thesis, University of Natural Resources and Life Sciences (BOKU), Vienna, Austria, 2014.

81. Melcher, A.; Ouédraogo, R.; Moog, O.; Slezak, G.; Savadogo, M.; Sendzimir, J. Healthy fisheries sustain society and ecology in Burkina Faso. In *Riverine Ecosystem Management*; Schmutz, S., Sendzimir, J., Eds.; Springer: Cham, Switzerland, 2018; pp. 519–539.

82. MRAH. *Stratégie Nationale De Développement Durable De La Pêche Et De l'Aquaculture à L'Horizon 2025*; Ministère des Ressources Animales et Halieutiques: Ouagadougou, Burkina Faso, 2013; p. 35.

83. FAO. *Code of Conduct for Responsible Fisheries*; FAO: Rome, Italy, 2019.

84. SUSFISH Consortium. *Sustainable Management of Water and Fish Resources in Burkina Faso-a Synthetic Overview of the SUSFISH Project. Supported by the Austrian Partnership Programme in Higher Education & Research for Development*; Final report; SUSFISH-Plus: Ouagadougou, Burkina Faso; Vienna, Austria, 2015; Available online: http://susfish.boku.ac.at/download.htm (accessed on 1 April 2019).

85. Sendzimir, J. Synthesis: Conceptual modelling. In *SUSFISH Book: Sustainable Fisheries and Water Management Transformation Pathways for Burkina Faso*, 1st ed.; Melcher, A.H., Ouedraogo, R., Oueda, A., Sendzimir, J., Slezak, G., Somda, J., Toé, P., Voigt, C., Eds.; SUSFISH-Plus: Ouagadougou, Burkina Faso; Vienna, Austria, 2020; in press.

86. Andrachuk, M.; Armitage, D. Understanding social-ecological change and transformation through community perceptions of system identity. *Ecol. Soc.* **2015**, *20*, 26. [CrossRef]

87. Ouedraogo, R.; Soara, A.E.; Oueda, A. Description du peuplement piscicole du lac sahélien de Higa, un site Ramsar du Burkina Faso, Afrique de l'Ouest. *J. Appl. Biosci.* **2015**, *95*, 8958–8965. [CrossRef]

88. Rösslberg, G. *Back to Basics: Traditional Inland Fisheries Management and Enhancement Systems in Sub-Saharan Africa and Their Potential for Development*; Deutsche Gesellschaftfür Technische Zusammenarbeit (GTZ) GmbH Eschborn: Eschborn, Germany, 2002.

89. Blaikie, P.; Brown, K.; Stocking, M.; Tang, L.; Dixon, P.; Sillitoe, P. Knowledge in action: Local knowledge as a development resource and barriers to its incorporation in natural resource research and development. *Agric. Syst.* **1997**, *55*, 217–237. [CrossRef]

90. Krupnik, I.; Jolly, D. *The Earth Is Faster Now: Indigenous Observations of Arctic Environmental Change*; Krupnik, I., Jolly, D., Eds.; Arctic Research Consortium of the United States: Fairbanks, AL, USA, 2002; ISBN 0-9720449-0-6.

91. Berkes, F.; Colding, J.; Folke, C. Rediscovery of traditional ecological knowledge as adaptive management. *Ecol. Appl.* **2000**, *10*, 1251–1262. [CrossRef]

92. Drew, J.A. Use of traditional ecological knowledge in marine conservation. *Conserv. Biol.* **2005**, *19*, 1286–1293. [CrossRef]

93. Fraser, D.J.; Coon, T.; Prince, M.R.; Dion, R.; Bernatchez, L. Integrating traditional and evolutionary knowledge in biodiversity conservation: A population level case study. *Ecol. Soc.* **2006**, *11*, 4. [CrossRef]

94. Sachs, J.; Warner, A.M. Sources of slow growth in African economies. *J. Afr. Econ.* **1997**, *6*, 335–376. [CrossRef]

95. Gilchrist, G.; Mallory, M.; Merkel, F. Can local ecological knowledge contribute to wildlife management? Case studies of migratory birds. *Ecol. Soc.* **2005**, *10*, 20. [CrossRef]

96. Ruddle, K.; Davis, A. What is "ecological" in local ecological knowledge? Lessons from Canada and Vietnam. *Soc. Nat. Resour.* **2011**, *24*, 887–901. [CrossRef]

97. Boonstra, W.; Nhung, P.H.T. The ghosts of fisheries management. *J. Nat. Resour. Policy Res.* **2011**, *4*, 1–25. [CrossRef]

98. Pokorny, B.; von Lübke, C.; Dayamba, S.D.; Dickow, H. All the gold for nothing? Impacts of mining on rural livelihoods in Northern Burkina Faso. *World Dev.* **2019**, *119*, 23–39. [CrossRef]

99. Bazillier, R.; Girard, V. The gold digger and the machine. Evidence on the distributive effect of the artisanal and industrial gold rushes in Burkina Faso. *J. Dev. Econ.* **2020**, *143*, 102411. [CrossRef]

100. Luxereau, A.; Genthon, P.; Karimou, J.M.A. Fluctuations in the size of Lake Chad: Consequences on the livelihoods of the riverain peoples in eastern Niger. *Reg. Environ. Chang.* **2012**, *12*, 507–521. [CrossRef]

101. Harrod, C. Climate change and freshwater fisheries. In *Freshwater Fisheries Ecology*; Craig, J.F., Ed.; John Wiley & Sons, Ltd.: Oxford, UK, 2015; pp. 641–694.

102. van de Giesen, N.; Liebe, J.; Jung, G. Adapting to climate change in the Volta Basin, West Africa. *Curr. Sci.* **2010**, *98*, 1033–1037.

103. Lebel, T.; Diedhiou, A.; Laurent, H. Seasonal cycle and interannual variability of the Sahelian rainfall at hydrological scales. *J. Geophys. Res. D Atmos.* **2003**, *108*, 1–11. [CrossRef]

104. IPCC. *Climate Change 2007: Contribution of Working Groups I, II and III to the Fourth Assessment Report of the Intergovernmental Panel on Climate Change*; The Core Writing Team, Pachauri, R.K., Reisinger, A., Eds.; IPCC: Geneva, Switzerland, 2007; ISBN 9291691224.

105. Bates, B.; Kundzewicz, Z.W.; Wu, S.; Palutikof, J. *Climate Change and Water. Technical Paper of the Intergovernmental Panel on Climate Change*; IPCC Secretariat: Geneva, Switzerland, 2008.

106. Paugy, D.; Lévêque, C.; Duponchelle, F. Life-history strategies. In *The Inland Water Fishes of Africa: Diversity, Ecology and Human Use*; Paugy, D., Lévêque, C., Otero, O., Eds.; IRD Éditions: Marseille, France, 2017; pp. 178–188.

107. Darwall, W.; Smith, K.; Allen, D.; Holland, R.; Harrison, I.; Brooks, E. *The Diversity of Life in African Freshwaters: Underwater, Under Threat: An Analysis of the Status and Distribution of Freshwater Species throughout Mainland Africa*; IUCN: Gland, Switzerland, 2011; ISBN 978-2-8317-1345-8.

108. Ouedraogo, R.; Soara, A.E.; Zerbo, H. Caractérisation du peuplement piscicole du réservoir de Boalin, Ziniaré (Burkina Faso) deux décennies après l'introduction de Heterotis niloticus. *Int. J. Biol. Chem. Sci.* **2015**, *9*, 2488–2499. [CrossRef]

109. Béné, C. *Diagnostic Study of the Volta Basin Fisheries: Part 1—Overview of the Fisheries Resources*; WorldFish Center Regional Offices for Africa and West Asia, Cairo Egypt, and CPWF: Colombo, Sri Lanka, 2007.

110. Béné, C.; Russell, A.J.M. *Diagnostic Study of the Volta Basin Fisheries: Part 2—Livelihoods and Poverty Analysis, Current Trends and Projections*; WorldFish Center Regional Offices for Africa and West Asia, Cairo, Egypt, and CPWF: Colombo, Sri Lanka, 2007.

111. Vye, S.R.; Emmerson, M.C.; Arenas, F.; Dick, J.T.A.; O'Connor, N.E. Stressor intensity determines antagonistic interactions between species invasion and multiple stressor effects on ecosystem functioning. *Oikos* **2015**, *124*, 1005–1012. [CrossRef]

112. Vinebrooke, R.D.; Kathryn, L.C.; Jon, N.; Marten, S.; Stanley, I.D.; Stephen, C.M.; Ulrich, S. Impacts of multiple stressors on biodiversity and ecosystem functioning: The role of species co-tolerance. *Oikos* **2004**, *104*, 451–457. [CrossRef]

113. Schinegger, R.; Palt, M.; Segurado, P.; Schmutz, S. Untangling the effects of multiple human stressors and their impacts on fish assemblages in European running waters. *Sci. Total Environ.* **2016**, *573*, 1079–1088. [CrossRef]

114. Nõges, P.; Argillier, C.; Borja, Á.; Garmendia, J.M.; Hanganu, J.; Kodeš, V.; Pletterbauer, F.; Sagouis, A.; Birk, S. Quantified biotic and abiotic responses to multiple stress in freshwater, marine and ground waters. *Sci. Total Environ.* **2016**, *540*, 43–52. [CrossRef]

115. Folt, C.L.; Chen, C.Y.; Moore, M.V. Synergism and antagonism among multiple stressors. *Limnol. Oceanogr.* **1999**, *44*, 864–877. [CrossRef]

116. Côté, I.M.; Darling, E.S.; Brown, C.J. Interactions among ecosystem stressors and their importance in conservation. *Proc. R. Soc. B Biol. Sci.* **2016**, *283*, 20152592. [CrossRef]

117. Przeslawski, R.; Byrne, M.; Mellin, C. A review and meta-analysis of the effects of multiple abiotic stressors on marine embryos and larvae. *Glob. Chang. Biol.* **2015**, *21*, 2122–2140. [CrossRef]

118. Harvey, B.P.; Gwynn-Jones, D.; Moore, P.J. Meta-analysis reveals complex marine biological responses to the interactive effects of ocean acidification and warming. *Ecol. Evol.* **2013**, *3*, 1016–1030. [CrossRef]

119. Vannevel, R. Using DPSIR and balances to support water governance. *Water* **2018**, *10*, 118. [CrossRef]

120. Patrício, J.; Elliott, M.; Mazik, K.; Papadopoulou, K.-N.; Smith, C.J. DPSIR-Two decades of trying to develop a unifying framework for marine environmental management? *Front. Mar. Sci.* **2016**, *3*, 177. [CrossRef]

121. Daniels, P.L. Climate change, economics and Buddhism—Part I: An integrated environmental analysis framework. *Ecol. Econ.* **2010**, *69*, 952–961. [CrossRef]

122. Nguyen, T.H.T.; Everaert, G.; Boets, P.; Forio, M.A.E.; Bennetsen, E.; Volk, M.; Hoang, T.H.T.; Goethals, P.L.M. Modelling tools to analyze and assess the ecological impact of hydropower dams. *Water* **2018**, *10*, 259. [CrossRef]

123. Melcher, A. *Annual Report of the SUSFISH-Plus Project*; University of Natural Resources and Life Sciences: Vienna, Austria, 2019.

124. Zan, L.M. Politiques de population et de réduction de la fécondité au Burkina Faso: Limites et perspectives. *Afr. Popul. Stud.* **2016**, *30*. [CrossRef]

125. Ministère de la Santé Plan. *National D'Accélération De Planification Familiale Du Burkina Faso 2017–2020*; Ministère de la Santé: Ouagadougou, Burkina Faso, 2017.

126. Anoh, A.; Fassassi, R.; Vimard, P. Politique de population et Planification Familiale en Côte d'Ivoire. In *Les Politiques De Planification Familiale: Cinq Expériences Nationales*; Gautier, A., Ed.; CEPED: Nogent-sur-Marne, France, 2004; pp. 195–231.

127. Breierova, L.; Duflo, E. *The Impact of Education on Fertility and Child Mortality: Do Fathers Really Matter Less Than Mothers?* Nber Working Paper Series; National Bureau of Economic Research: Cambridge, MA, USA, 2004.

128. Canning, D.; Raja, S.; Yazbeck, A.S. *Africa's Demographic Transition: Dividend or Disaster?* Africa Development Forum series; The World Bank: Washington, DC, USA, 2015; ISBN 978-1-4648-0490-8.

129. Toé, P. Pêche, environnement et société: Contribution des sciences sociales sur les pêcheries traditionnelles en pays Bisa (Burkina Faso). *Cah. du Cent. d'études Rech. en lettres Sci. Hum. Soc.* **1999**, *16*, 301–318.

130. Toé, P.; Sanon, V.-P. *Gouvernance Et Institutions Traditionnelles Dans Les Pêcheries De l'Ouest Du Burkina Faso*; L'Harmattan: Paris, France, 2015; ISBN 978-2-343-04967-0.

131. Buur, L.; Kyed, H.M. *State Recognition and Democratization in Sub-Saharan Africa*; Palgrave Macmillan: New York, NY, USA, 2007; ISBN 978-0-230-60971-6.

132. Froehlich, H.E.; Gentry, R.R.; Halpern, B.S. Conservation aquaculture: Shifting the narrative and paradigm of aquaculture's role in resource management. *Biol. Conserv.* **2017**, *215*, 162–168. [CrossRef]

133. De Graaf, G. Sustainable Fisheries Management and Culture-Based Fisheries in Reservoirs a Case Study from Burkina Faso. 2003. Available online: http://citeseerx.ist.psu.edu/viewdoc/download?doi=10.1.1.534.6168&rep=rep1&type=pdf (accessed on 1 April 2019).

134. Allison, E.H.; Ellis, F. The livelihoods approach and management of small-scale fisheries. *Mar. Policy* **2001**, *25*, 377–388. [CrossRef]

135. Kaboré, I.; Moog, O.; Ouéda, A.; Sendzimir, J.; Ouedraogo, R.; Guenda, W.; Melcher, A.H. Developing reference criteria for the ecological status of West African rivers. *Environ. Monit. Assess.* **2018**, *190*, 2. [CrossRef]

136. Zerbo, H.; Ouattara, D.; Soubeiga, Z.; Kabore, K.; Kabore, C.; Bado, E.; Goumbri, B.A.; Yerbanga, R.A.; Ouedraogo, N.; Baro, S. *Analyse De La Filière Pêche Au Burkina Faso. Projet d'Appui Au Renforcement des Capacités d'Analyse Des Impacts Des Politiques Agricoles*; Direction Générale des Ressources Halieutiques. Ministère de l'Agriculture, de l'Hydraulique et des Ressources Halieutiques: Ouagadougou, Burkina Faso, 2007.

137. Miller, J. The potential for development of aquaculture and its integration with irrigation within the context of the FAO special programme for food security in the Sahel. In *Integrated Irrigation and Aquaculture in West Africa: Concepts, Practices and Potential*; Halwart, M., Dam, A.A., Eds.; Food and Agriculture Organization of the United Nations: Rome, Italy, 2006.

138. Obiero, K.; Meulenbroek, P.; Drexler, S.; Dagne, A.; Akoll, P.; Odong, R.; Kaunda-Arara, B.; Waidbacher, H. The contribution of fish to food and nutrition security in Eastern Africa: Emerging trends and future outlooks. *Sustainability* **2019**, *11*, 1636. [CrossRef]

139. Iwama, G.K. Interactions between aquaculture and the environment. *Crit. Rev. Environ. Control* **1991**, *21*, 177–216. [CrossRef]

140. George, A.L.; Kuhajda, B.R.; Williams, J.D.; Cantrell, M.A.; Rakes, P.L.; Shute, J.R. Guidelines for propagation and translocation for freshwater fish conservation. *Fisheries* **2009**, *34*, 529–545. [CrossRef]

141. Fraser, D.J. How well can captive breeding programs conserve biodiversity? A review of salmonids. *Evol. Appl.* **2008**, *1*, 535–586. [CrossRef] [PubMed]

Review

Agricultural Effects on Streams and Rivers: A Western USA Focus

Robert M. Hughes [1,2,]* and Robert L. Vadas, Jr. [3]

[1] Amnis Opes Institute, Corvallis, OR 97333, USA
[2] Department of Fisheries, Wildlife, and Conservation Sciences, Oregon State University, Corvallis, OR 97331, USA
[3] 2909 Boulevard Road Southeast, Olympia, WA 98501, USA; bobesan@comcast.net
* Correspondence: hughes.bob@amnisopes.com

Abstract: Globally, croplands and rangelands are major land uses and they have altered lands and waters for millennia. This continues to be the case throughout the USA, despite substantial improvements in treating wastewaters from point sources—versus non-point (diffuse) sources. Poor macroinvertebrate assemblage condition occurs in 30% of conterminous USA streams and rivers; poor fish assemblage condition occurs in 26%. The risk of poor fish assemblage condition was most strongly associated with excess nutrients, salinity and sedimentation and impaired riparian woody vegetation. Although the Clean Water Act was passed to restore and maintain the integrity of USA waters, that will be impossible without controlling agricultural pollution. Likewise, the Federal Land Policy and Management Act was enacted to protect the natural condition of public lands and waters, including fish habitat, but it has failed to curtail the sacred cows of livestock grazing. Although progress has been slow and spotty, promising results have been obtained from basin and watershed planning and riparian zone protections.

Keywords: USA; fish assemblages; macroinvertebrate assemblages; bird assemblages; croplands; rangelands

Citation: Hughes, R.M.; Vadas, R.L., Jr. Agricultural Effects on Streams and Rivers: A Western USA Focus. *Water* **2021**, *13*, 1901. https://doi.org/10.3390/w13141901

Academic Editors: Pedro Segurado, Paulo Branco and Maria Teresa Ferreira

Received: 17 May 2021
Accepted: 5 July 2021
Published: 9 July 2021

Publisher's Note: MDPI stays neutral with regard to jurisdictional claims in published maps and institutional affiliations.

1. Introduction

1.1. What Is the Biological Condition of All USA Streams and Rivers?

The latest national assessment of conterminous USA streams and rivers indicated that only 26–30% of the entire stream/river length was in good condition based on samples of 1924 randomly selected sites [1]. For macroinvertebrate assemblage condition determined from macroinvertebrate multimetric index (MMI) scores, it was 30% nationally (22% and 51% in the Xeric and Western Mountains ecoregions, respectively; Figure 1). For fish assemblage MMIs, those numbers were 26% nationally, 26% Western Mountains, and 19% Xeric. Nationally and west-wide, 4–58% of the stream/river length was in poor condition for total phosphorus, total nitrogen, riparian woody vegetation and riparian disturbance. The relative risk of poor fish assemblage MMI scores nationally, given a poor stressor score, was greatest for total phosphorus, total nitrogen, riparian woody vegetation, excess sedimentation and excess salinity. In the western USA, the greatest relative risks for poor fish assemblage MMI scores were for poor riparian woody vegetation, excess fine sediments, and excess salinity [2]. Using logistic regression analysis, Herlihy et al. [3] determined that poor fish MMI scores were 106 and 20.6 times as likely to occur as a result of excess salinity and excess fine sediment in the Xeric and Mountains Ecoregions of the western USA, respectively. In general, most variables for predicting both fish and macroinvertebrate MMI scores were local site variables (e.g., water quality and substrate size). However, dam density was also important for macroinvertebrates in the Xeric and Mountains Ecoregions, whereas catchment development was important for fish in the Mountains Ecoregion. Thus, a considerable proportion of USA and western USA stream length is in poor condition because of poor water quality, excess sedimentation, and degraded woody riparian vegetation.

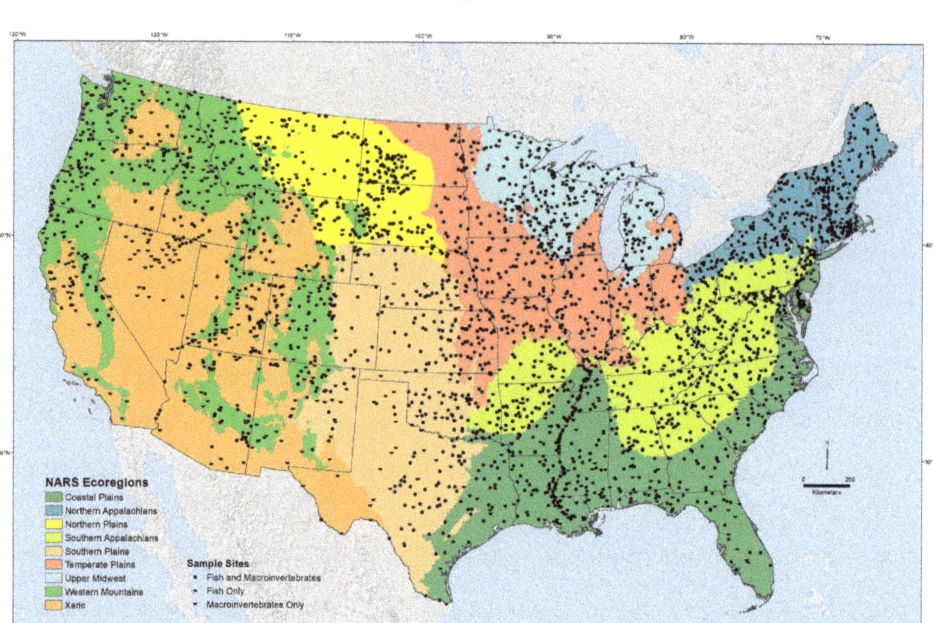

Figure 1. Locations of the NRSA sample sites and the nine aggregate ecoregions used for pattern analysis (from USEPA 2020).

1.2. What Is the Major Anthropogenic Pressure on Streams?

Nationally, agriculture was deemed the cause of 48% of water-quality impairment in USA surface waters [4]. Based on nonmetric multidimensional scaling of Bray–Curtis similarity analyses, Brown et al. [5] determined that prior agricultural or forest land use was the most important factor affecting correlations between fish and macroinvertebrate assemblages along an urbanization intensity gradient in urban streams across nine USA metropolitan areas. Chen and Olden [6], using gradient forest modeling, determined threshold changes in fish species richness and assemblage composition at 26% and 31% of catchment agriculture, respectively, for conterminous USA hydrologic units. Clearly, agricultural land uses are driving poorer water quality and poorer aquatic biotic conditions nationally.

Similar patterns are evident in the western USA. USDI [7] found that 66–78% of the riparian zones in western rangelands were damaged by livestock grazing and were in their worst condition in history. The percentage of total catchment as irrigated agriculture explained 56% of fish assemblage MMI scores, which declined with increased agriculture, in Pacific Northwest rivers [8]. Carlisle and Hawkins [9] found that macroinvertebrate assemblage condition scores were significantly lower for farm and rangeland (grazed) sites than forested sites in the western USA. Mulvey et al. [10] reported that agricultural lands accounted for 80% of the impaired stream length in the Willamette Basin, Oregon, despite representing only 30% of total stream length. Riseng et al. [11], using structural equation modeling (SEM), determined that percent catchment agriculture in the Columbia Plateau and Upper Snake River hydrologic units increased temperature and reduced flow, coarse substrate, and macroinvertebrate MMI scores. Beschta et al. [12] reported that livestock altered 939,000 km^2 of western USA public lands, over an order of magnitude more than is altered by roads, fire and logging combined. Perkin et al. [13], also employing SEM, determined that total catchment agriculture was associated with reduced stream fish richness in Great Plains streams. Using random forest modeling, Hill et al. [14] determined that the most important anthropogenic predictors of macroinvertebrate MMI scores were urbanization and agriculture both nationally and in the Xeric and Mountains Ecoregions. Perkin et al. [15] reported that water diversions and aquifer pumping in the Great Plains were associated with fragmented streams and loss of 558 stream km, which in turn transformed fish assemblages from dominance by large-stream fishes to small-stream fishes. Saunders and Fausch [16] determined that riparian-derived prey in trout diets was reduced by 51–74% at increasing levels of livestock grazing compared against exclosures. Jacobson et al. [17] found that eutrophication from agriculture was the major stressor of coldwater fish habitat in Eastern Temperate Forests and Great Plains Ecoregion lakes. Based on multiple regression modeling, Kaufmann et al. (unpublished data, USEPA, Corvallis, OR, USA) determined that riparian and catchment agriculture were the most important anthropogenic pressures associated with poor streambed stability, woody riparian vegetation condition, and fish habitat cover across the USA and in the Western Mountain and Xeric Ecoregions. Therefore, agricultural land uses (including livestock grazing) are driving poorer water quality and aquatic biotic conditions in the western USA.

1.3. What Happens When Forests Are Converted to Agriculture?

Such patterns as those discussed above are particularly evident when forested regions are converted to agriculture. Leitão et al. [18], using SEM, determined that local and catchment deforestation decreased instream large wood, which reduced fish species richness and functional originality (uniqueness) in one region but not another. Percent pasture or percent agriculture were the major land uses associated with poor macroinvertebrate MMI scores in several river basins [19,20]. Threshold indicator taxa analyses revealed thresholds at 1–12% riparian forest loss and 9% total catchment forest loss for macroinvertebrate taxa [21,22], and 6–10% riparian forest loss and 1–10% total catchment forest loss for fish species [23,24]. As indicated above, even very low levels of forest and savanna devegetation can lead to the extirpation of sensitive species, which likely happened a century or more ago in the USA.

The importance of agriculture and livestock grazing to poor stream conditions is a function of three major factors. (1) Agriculture is one of the most widespread and intensive land uses (17% of the conterminous USA land area; 8% of western USA land area) [25]. (2) Rangelands are more extensive, comprising 29% of conterminous USA land area, mostly in the western states [25]. (3) Croplands are even more poorly regulated and more intensively altered than the other two major western and USA land uses: forestlands and rangelands. Thus, it is no wonder that agriculture and livestock grazing drive impairments of most USA stream kilometers—but what can be done about it?

2. Case Studies

2.1. Cropland Case Studies: Research and Management Implications

We conducted a literature search to locate at least 20 case studies each that related (1) agricultural best management practices (BMPs) to instream aquatic biotic responses and (2) livestock exclosures to instream and riparian faunal responses. As expected, most of the former studies were located in the agricultural Midwest and Southeast USA states, where row crop agriculture predominates (Figure 2). On the other hand, most of the exclosure studies were located in the western USA states where rangelands and livestock grazing predominate. Nearly half of the agricultural BMP studies involved fewer than 20 sites; 75% of the exclosure studies involved fewer than 20 sites. Only one study effectively calibrated for natural differences in catchment geology and geomorphology [26]. None of the 44 studies incorporated a probability survey design; in addition, the sampling methods and indicators were inconsistent among the studies. These sample sizes, methods, and survey constraints limit the degree to which the study results can be inferred confidently to the USA or any USA state or region [27,28]. Nonetheless, we found that several important patterns emerged from catchment and riparian BMP studies, as listed below (Table 1).

Table 1. Case studies of the effects of improved agricultural management on stream biota.

State or Region	Study Design	Sites	Mgmt. Practice	Indicators	Results	Source
Wisconsin	disturbance gradient	25	conversion of farmland to forest	fish, diatom and macroinvertebrate MMIs	increased MMI scores	[26]
North Carolina	disturbance gradient	3	conversion of farmland to forest	fish and macroinvertebrate MMIs	increased MMI scores	[29]
Michigan	disturbance gradient	23	conversion of farmland to forest	fish MMI	increased MMI scores	[30]
Wisconsin	disturbance gradient	134	conversion of unwooded to wooded riparian zones and catchments	Fish MMI	Increased scores	[31]
Minnesota	disturbance gradient	20	conversion of unwooded to wooded riparian zones	fish MMI	increased MMI scores	[32]
Michigan	disturbance gradient	23	conversion of unwooded to wooded riparian zones and catchments	fish MMI	increased MMI scores, especially for catchments	[33]
Wisconsin	disturbance gradient	38	conversion of unwooded to wooded riparian zones and catchments	fish and macroinvertebrate MMIs	increased MMI scores	[34]
Illinois	disturbance gradient	84	remove agricultural land from production	EPT taxa richness	no effect	[35]
Minnesota	disturbance gradient	3	agricultural land retirement	fish MMI	improved with riparian agricultural retirement	[36]
Missouri basin	disturbance gradient	526	conservation practices	lithophilic fish	>50% land treatment to have significant effect	[37]
North Carolina	disturbance gradient	3	erosion control	Ephemeroptera Plecoptera Trichoptera	increased taxa and EPT richness	[38]
Missouri and Arkansas	disturbance gradient	30	reduced livestock production	fish, diatom and macroinvertebrate MMIs	increased MMI scores	[39]
USA	disturbance gradient	172	conversion of unwooded to wooded riparian zones	fish MMI	increased MMI scores	[40]
Minnesota	disturbance gradient	20	conversion of unwooded to wooded riparian zones	fish MMI	Increased MMI scores	[41]
Indiana	before-after	2	re-meandering	fish	minimal and negative effects	[42]
North Carolina and Virginia	disturbance gradient	3	livestock exclusion; channel rehabilitation; agriculture BMPs	macroinvertebrates	conditions declined in 2 sites and improved in the BMP site	[43]
Wisconsin	BACI	4	agriculture BMPs	fish assemblage	improved in 1 BMP site	[44]
Ohio	BACI	16	no-till and low-till agriculture	fish MMI	significantly improved MMI scores	[45]
Illinois	disturbance gradient	9	wooded riparian buffers	fish and macroinvertebrates	abundances decreased and fish MMI scores increased	[46]
Virginia	paired	48	riparian buffers	fish MMI	scores increased	[47]
Georgia	paired	5	riparian buffers	macroinvertebrates and amphibians	scores increased	[48]

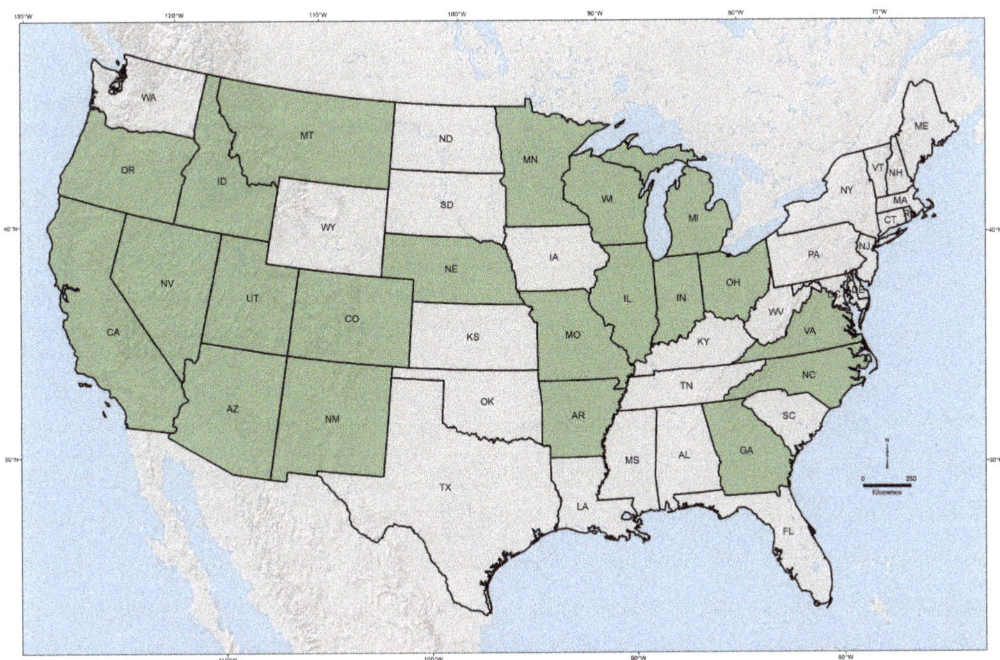

Figure 2. Locations of case study states (shaded). State abbreviations: AR (Arkansas), AZ (Arizona), CA (California), CO (Colorado), GA (Georgia), ID (Idaho), IL (Illinois), IN (Indiana), MI (Michigan), MN (Minnesota), MO (Missouri), MT (Montana), NE (Nebraska), NC (North Carolina), NM (New Mexico), NV (Nevada), OH (Ohio), OR (Oregon), UT (Utah), VA (Virginia), WI (Wisconsin).

- Both catchment and riparian treatments can affect site MMI scores [34], with the degree of those effects being a function of the relative degrees of disturbance at those two spatial extents. Where catchment conditions are intensively and extensively altered, site-specific BMPs have limited effectiveness. Where this is not the case, site-specific BMPs can produce significant improvements [41]. In other words, riparian BMPs can improve site habitat conditions, but fish assemblages cannot be recovered if there is insufficient catchment BMP implementation [31,49]. Thus, study extents matter.
- Biotic relationships with agricultural land use are very complex. Clear increases in MMI scores were apparent only after agricultural land use was less than 50%. However, even with 80% agricultural land use, some sites with relatively high gradients and rocky substrate that had not been channelized had high MMI scores [31].
- Together with historical land and water uses, unanticipated land disturbances and BMPs occurred during studies, thereby confounding the results of both BACI and disturbance gradient studies [43].
- Study durations were often insufficient to detect changes resulting from agricultural BMPs and stream-channel rehabilitation [42,43].
- Contrasting results, even from studies in the same river basin, occur because of the differing spatial extents of their study designs, together with the strengths of the relationships between stream biotic conditions and the differing effectiveness of the catchment and riparian BMP treatments expected to affect those conditions [30].
- In the Midwest, both grass and wood riparian buffers improved macroinvertebrate and fish indicator scores [50]. Therefore, it is important to consider the potential natural vegetation of riparian buffer zones rather than always planting trees (especially non-native species).

- Different indicators have different sensitivities to the same pressures or stressors [38,39]; different sensitivities to different pressures and stressors [26]; and differing sensitivities at catchment, riparian corridor, and site extents [34].
- Total taxa richness is an illusionary indicator when sensitive taxa are replaced by tolerant taxa [38]. Moreover, fish and macroinvertebrate taxa richness estimates are strongly affected by sample size and sampling effort [51–53].
- Total abundance often indicates nutrient enrichment of streams [38].

2.2. Livestock Exclosure Case Studies: Research and Management Implications

Livestock exclosure studies have many of the same constraints as catchment/riparian BMP studies that are listed above, plus others that are unique to riparian exclosures and the rangelands where most exclosure studies were located. However, we found that several important patterns emerged from catchment and riparian BMP studies, as listed below (Table 2).

Table 2. Case studies of the effects of livestock exclosures on stream/riparian fauna.

State or Region	Study Design	Sites	Indicators	Results	Source
Minnesota	disturbance gradient	17	fish and macroinvertebrates	varied more by buffer type than grazing intensity	[50]
Nebraska	disturbance gradient	6	macroinvertebrate MMI	improved scores	[54]
New Mexico	paired	4	tolerant macroinvertebrates	decreased densities and biomasses	[55]
California	paired	38	macroinvertebrates	richness increased	[56]
Oregon	paired	9	macroinvertebrates	abundance increased	[57]
Virginia	paired	10	macroinvertebrates	no significant difference	[58]
Wisconsin	paired	16	macroinvertebrates	improved scores	[59]
Minnesota	paired	26	macroinvertebrate MMI	improved scores	[60]
Oregon	paired	16	fish	increased age-0 Redband Trout densities	[61]
California	paired	7	Golden Trout	increased density and biomass	[62]
Oregon, Utah, Montana	paired	10	trout biomass	increased 184%	[63]
Idaho	paired	6	trout	abundance and size increased	[64]
Colorado	paired	3	trout biomass	doubled	[65]
Arizona	paired	6	riparian birds	increased density and species richness	[66]
Idaho	BACI	14	fish and macroinvertebrates	increased age-0 salmonid densities	[67]
Oregon	BACI	69	riparian birds	increased abundance and richness of species of concern	[68]
Oregon	BACI	106	riparian birda	increased abundance and richness	[69]
California, Idaho, Montana, Nevada, Oregon	BACI	437	riparian birds	increased abundance and richness	[70]
Oregon	BACI	9	riparian birds	increased abundance and richness	[71]
Oregon	BACI	6	riparian birds	increased abundance and richness	[72]

- Proximate paired sites on the same streams typically are not independent; rather they tend to be pseudoreplicates [73], meaning that upstream conditions may have important biological effects on downstream conditions in an exclosure, and vice versa. Both conditions confound biological responses to exclosures [58].

- Small natural differences in channel slope, morphology and substrate may confound comparisons between the instream biological effects of exclosures versus grazed riparian zones [56].
- As with catchment versus riparian agriculture, small-sized and short-term grazing exclosures tend to be less effective measures for recovering aquatic biotic condition than livestock removal at larger spatial extents [57,59].
- Even more so than agricultural BMPs, exclosure projects have been ad hoc, not selected as part of long-term survey designs and lacking controls that could be tested efficiently [61].
- Most exclosures are too short spatiotemporally to reduce fine sediment loads and summer water temperatures sufficiently, let alone be sufficient to incorporate the riverscapes that salmonids require to successfully complete their life histories over multiple seasons and years [74,75].
- Total abundance of riparian birds frequently indicates catchment disturbance that increases abundances of wide-ranging generalist taxa [70].
- Although both macroinvertebrate and fish indicators usually had improved scores inside livestock exclosures, those responses for riparian birds tended to be stronger and more consistent (Table 2). Presumably, this occurred because of the stronger relationship between riparian vegetation and bird assemblages, and the longer durations of riparian recoveries in the avian studies.

3. Discussion

3.1. Major BMP Research and Management Challenges

There are many challenges in planning, implementing, and monitoring spatially extensive programs for improving the biotic condition of streams draining croplands and rangelands.

- Holistic, basin-extent plans for implementing and monitoring rehabilitation projects are lacking [76].
- Planning—and its monitoring and indicators—must include the geographic context and be implemented at appropriate spatial extents [76,77].
- Targeted approaches addressing entire stream lengths and their associated catchments are required to restore aquatic ecosystem integrity given the pervasive effects of croplands and overgrazing on riverscapes. Overgrazing and farming limit the degree to which significant proportions of stream networks can be rehabilitated [58,60,67]. Therefore, BMPs of multiple types should be aggregated in catchments and in proximity to streams and their floodplains to maximize effectiveness, and those BMPs must be maintained [77].
- The monitoring and indicators must be linked to specific objectives and predicted ecosystem improvements [76], and it is critical to collect quantitative pre- and post-BMP water quality, physical habitat structure and biological data, including multiple indicators for each [77].
- The survey designs, monitoring protocols, indicators and funding must be commensurate with the extent of the problem [76].
- The planning, rehabilitation and monitoring must be collaborative—not limited and parochial [76].
- That collaboration must extend to employing multiple indicators, particularly riparian birds, when assessing the effects of riparian buffers and livestock exclosures, as has been observed in lake studies [78–80].
- Greater collaboration must occur among landowners and local, state, and federal agencies that regulate land and water management in river basins, because local agencies typically lack the knowledge and authority to holistically govern up- and downstream conditions [30,81].

- Historical land uses and time lags following project implementation must be incorporated into project planning and monitoring [43,77,82]. For example, time lags following historical or current land-use changes, particularly their effects on nutrient residence times in groundwater, mean that decades are required to remove them from agricultural groundwaters feeding streams. Similarly, fine sediments and phosphorus move slowly through river networks because of storage and remobilization processes, especially in low-slope agricultural streams, where their removal may require decades to centuries [82].
- Planning for the thermal and hydrological impacts of current and future climate change is essential [77], particularly the increasing likelihood of extreme weather events, such as floods, droughts, fire and high winds.
- Livestock exclosure and stream-rehabilitation research has produced considerable scientific uncertainty because of relatively few studies, weak study designs and indicators, and insufficient consideration of the spatial extents and mechanisms of ecosystem recovery [83]. Exclosure and rehabilitation projects are generally too small and poorly located to measure aquatic indicator responses to livestock removal or BMPs accurately and precisely. Project response timing and dynamics may vary considerably with location and treatment. Sites can recover relatively quickly and predictably, recover slowly and remain more sensitive to impacts than they were before project initiation, or fail to recover at all.
- The scientific foundations for livestock exclosure and stream rehabilitation research can be improved by developing long-term, spatially extensive research programs; better project placement and study designs; and stronger commitments to pretreatment data collection [76,77,83,84].
- By altering stream catchments, humans degrade stream/riparian ecosystems in multiple ways [85]. However, fully understanding the relationships between land/stream uses and stream ecological condition is complicated by the covariation of anthropogenic and natural gradients, the differing effects of different spatial extents, and uncertainties surrounding the importance of land use legacies, physicochemical and biotic indicator sensitivities, and those indicator response thresholds [22,85–88].
- The most critical step in stream rehabilitation is cessation of the anthropogenic activities that cause degradation and hinder recovery [89]. Before implementing active rehabilitation projects, allowing sufficient time for natural recovery is recommended. Not doing so can actually exacerbate the degree of degradation and further hinder rehabilitation. Rehabilitation should be focused initially on catchments rather than riparian/stream ecosystems, assuming the catchments and their floodplains are driving degraded stream conditions [85,90].
- For projects focused on riparian zones, establish them as separate management units with different management objectives than their catchments. Limit livestock by herding, controlling the timing, intensity and duration of grazing, or permanently fencing them off from grazing. Limit agriculture to allow the potential natural riparian and floodplain vegetation to recover and monitor land use for compliance. At least on public lands, establish grazing and cropland fees commensurate with the costs of management and monitoring [91].
- Stream riparian buffer management offers largely extent-independent effects (shading, thermal controls, and organic matter and large wood additions) [92]. However, catchment management offers extent-dependent effects (nutrients and fine sediment retention, as well as flow regime) [92]. Extent-dependent effects and variations in riparian management often limit the biological responses of local riparian management. Concerted management across both spatial extents is required for full biological recovery of damaged streams. Nonetheless, the ecological benefits of wide riparian buffers along entire channel networks outweigh any potential adverse ecological effects, particularly for small streams [77,92].

3.2. What Can Be Done to Reduce Agricultural Impacts on Streams?

The science is clear. Although the objective of the Clean Water Act (CWA) is to "restore and maintain the chemical, physical and biological integrity of the Nation's waters," that will be impossible without controlling agricultural pollution. There are at least four key reasons why USA agriculture is so inadequately regulated under the CWA) [93]. (1) Irrigated agriculture and agricultural drains are explicitly exempted from federal discharge permitting. (2) Pollutant discharges are restricted to point sources (usually pipes) versus diffuse sources, which are delegated to the states or local jurisdictions to manage via so-called BMPs. (3) Although CWA Sections 208 and 319 ask states to adopt basin-wide land use plans to control diffuse pollution, USEPA lacks the authority to determine the adequacy of those plans or to develop alternative plans, unlike what it does for point sources. In addition, local governments and landowners have resisted land use controls, federal funding for 208 planning ended in 1981, and Congressional funding for Section 319 planning and implementation is insufficient. (4) Despite its objectives, CWA enforcement still focuses mostly on water quality pollution, ignoring the substantial impacts of agriculture on physical habitat structure, flow regimes, riparian zones and biota [94–97]. For example, Ohio EPA [98] detected stream biological impairment 50% of the time that chemical criteria were met. This means that other stressors, such as physical habitat structure, limit biological condition and that biological assessments are more sensitive to landscape pressures and local stressors than are chemical toxicity criteria [1,3].

3.3. What USA Policies Might Be Implemented to Reduce Agricultural Impacts on Streams?

There are several options for how the USEPA and state and local agencies can mitigate the problems of diffuse pollution from croplands and rangelands (as well as forestlands). Under CWA Section 303, USEPA can impose water body standards (which include designated uses and the environmental criteria needed to protect those uses) on states that fail to implement those standards. However, the federal government has been reluctant to enforce that law because of its implications regarding the property rights of millions of farmers and ranchers. That Section also requires that states identify and list impaired waters that fail to meet standards and then abate that pollution, whether from point or diffuse sources by establishing total maximum daily loads (TMDLs) of the limiting pollutants. To limit those pollutants, farmers, ranchers and other landowners will need to limit nutrient, biocide and sediment runoff—most likely by limiting soil erosion and restoring riparian vegetation buffers. However, federal courts have been inconsistent in supporting such controls on pollution [93]. Furthermore, the U.S. Department of Agriculture's Conservation Reserve Program pays farmers to remove ecologically sensitive cropland for 10–15 years, amounting to millions of protected hectares per year. Nonetheless, when contracts expire about half the land is returned to crop production [99]. The Federal Land Planning and Management Act of 1975 requires that public lands (and their waters) be managed for sustainable use, including protection of their natural condition (where appropriate), and provide food and habitat for fish and wildlife [100]. As indicated by the conditions of waters draining western USA rangelands summarized above, these requirements are infrequently met and depend on supportive federal courts for ensuring accountability (e.g., [101,102]). Increasingly, basin and watershed councils have reached consensus among landowners to implement basin- and watershed-wide management plans and TMDLs that encompass both point and diffuse sources [103–105]. In both Europe and the USA, many stream rehabilitation projects have focused on riparian protections, but few of them have been evaluated rigorously for instream effectiveness [83,106,107]. This includes a need to focus on biotic and groundwater variables, rather than just surface water-quality parameters that tend to overestimate riparian-buffer effectiveness for aquatic-ecosystem protection [83,96,108]. Clearly, if entire catchments are converted to intensive agriculture or livestock grazing, the potentials for obtaining good stream biological conditions are limited [76,77,108]. Nonetheless, protection and rehabilitation of riparian zones can increase the probability of improved biological status in many cases [16,76,92,109,110]. Policies that encourage doing so—and discourage not doing so–are warranted.

Author Contributions: Conceptualization R.M.H.; writing—original draft R.M.H.; writing—review and editing R.M.H. & R.L.V.J. Both authors have read and agreed to the published version of the manuscript.

Funding: This research received no external funding.

Acknowledgments: We thank Randy Comeleo for the two maps and Eric Featherman and two anonymous reviewers for helpful comments on an earlier manuscript. We also appreciate the invitation to contribute the manuscript from Special Issue Editors Pedro Segurado, Teresa Ferreira and Paulo Branco.

Conflicts of Interest: We declare no conflict of interest. Our employers had no role in writing or publishing the manuscript.

References

1. USEPA (U.S. Environmental Protection Agency). *National Rivers and Streams Assessment 2013–2014: A Collaborative Survey*; EPA 841-R-19-001; Office of Water and Office of Research and Development: Washington, DC, USA, 2020. Available online: https://www.epa.gov/national-aquatic-resource-surveys/nrsa (accessed on 6 July 2021).
2. USEPA (U.S. Environmental Protection Agency). *National Rivers and Streams Assessment 2008–2009: A Collaborative Survey*; EPA/841/R-16/007; Office of Water/Office of Research and Development: Washington, DC, USA, 2016. Available online: https://www.epa.gov/sites/production/files/2016-03/documents/nrsa_0809_march_2_final.pdf (accessed on 6 July 2021).
3. Herlihy, A.T.; Sifneos, J.C.; Hughes, R.M.; Peck, D.V.; Mitchell, R.M. Relation of lotic fish and benthic macroinvertebrate condition indices to environmental factors across the conterminous USA. *Ecol. Indic.* **2020**, *112*, 105958. [CrossRef]
4. USEPA (U.S. Environmental Protection Agency). National Water Quality Inventory Report. EPA-841-F-02-003; 2002. Available online: https://www.epa.gov/sites/production/files/2015-09/documents/2007_10_15_305b_2002report_report2002305b.pdf (accessed on 6 July 2021).
5. Brown, L.R.; Cuffney, T.F.; Coles, J.F.; Fitzpatrick, F.; McMahon, G.; Steuer, J.; Bell, A.H.; May, J.T. Urban streams across the USA: Lessons learned from studies in 9 metropolitan areas. *J. N. Am. Benthol. Soc.* **2009**, *28*, 1051–1069. [CrossRef]
6. Chen, K.; Olden, J.D. Threshold responses of riverine fish communities to land use conversion across regions of the world. *Glob. Chang. Biol.* **2020**, *26*, 4952–4965. [CrossRef]
7. USDI (U.S. Department of the Interior). *Rangeland Reform '94: Draft Environmental Impact Statement*; Bureau of Land Management: Washington, DC, USA, 1994.
8. Mebane, C.A.; Maret, T.R.; Hughes, R.M. An index of biological integrity (IBI) for Pacific Northwest rivers. *Trans. Am. Fish. Soc.* **2003**, *132*, 239–261. [CrossRef]
9. Carlisle, D.M.; Hawkins, C.P. Land use and the structure of western US stream invertebrate assemblages: Predictive models and ecological traits. *J. N. Am. Benthol. Soc.* **2008**, *27*, 986–999. [CrossRef]
10. Mulvey, M.; Leferink, R.; Borisenko, A. *Willamette Basin Rivers and Streams Assessment*; Oregon Department of Environmental Quality: Hillsboro, OR, USA, 2009.
11. Riseng, C.M.; Wiley, M.J.; Black, R.W.; Munn, M.D. Impacts of agricultural land use on biological integrity: A causal analysis. *Ecol. Appl.* **2011**, *21*, 3128–3146. [CrossRef]
12. Beschta, R.L.; Donahue, D.L.; DellaSala, D.A.; Rhodes, J.J.; Karr, J.R.; O'Brien, M.H.; Fleischner, T.L.; Williams, C.D. Adapting to climate change on western public lands: Addressing the ecological effects of domestic, wild, and feral ungulates. *Environ. Manag.* **2013**, *51*, 474–491. [CrossRef] [PubMed]
13. Perkin, J.S.; Troia, M.J.; Shaw, D.C.R.; Gerken, J.E.; Gido, K.B. Multiple watershed alterations influence community structure in Great Plains prairie streams. *Ecol. Freshw. Fish.* **2014**, *25*, 141–155. [CrossRef]
14. Hill, R.A.; Fox, E.W.; Leibowitz, S.G.; Olsen, A.R.; Thornbrugh, D.J.; Weber, M.H. Predictive mapping of the biotic condition of conterminous U.S. rivers and streams. *Ecol. Appl.* **2017**, *27*, 2397–2415. [CrossRef] [PubMed]
15. Perkin, J.S.; Gido, K.B.; Falke, J.A.; Fausch, K.D.; Crockett, H.; Johnson, E.R.; Sanderson, J. Groundwater declines are linked to changes in Great Plains stream fish assemblages. *Proc. Natl. Acad. Sci. USA* **2017**, *114*, 7373–7378. [CrossRef] [PubMed]
16. Saunders, W.C.; Fausch, K.D. Conserving fluxes of terrestrial invertebrates to trout in streams: A first field experiment on the effects of cattle grazing. *Aquat. Conserv. Mar. Freshw. Ecosyst.* **2019**, *28*, 910–922. [CrossRef]
17. Jacobson, P.C.; Hansen, G.J.A.; Olmanson, L.G.; Wehrly, K.E.; Hein, C.L.; Johnson, L.B. Loss of coldwater fish habitat in glaciated lakes of the midwestern United States after a century of land use and climate change. *Am. Fish. Soc. Sympos.* **2019**, *90*, 141–157.
18. Leitão, R.P.; Zuanon, J.; Mouillot, D.; Leal, C.G.; Hughes, R.M.; Kaufmann, P.R.; Villéger, S.; Pompeu, P.S.; Kasper, D.; de Paula, F.R.; et al. Disentangling the pathways of land use impacts on the functional structure of fish assemblages in Amazon streams. *Ecography* **2018**, *41*, 219–232. [CrossRef] [PubMed]

19. Silva, D.R.O.; Herlihy, A.T.; Hughes, R.M.; Macedo, D.R.; Callisto, M. Assessing the extent and relative risk of aquatic stressors on stream macroinvertebrate assemblages in the neotropical savanna. *Sci. Tot. Environ.* **2018**, *633*, 179–188. [CrossRef]

20. Martins, I.; Macedo, D.R.; Hughes, R.M.; Callisto, M. Major risks to aquatic biotic condition in a Neotropical Savanna river basin. *River Res. Appl.* **2021**, *37*, 858–868. [CrossRef]

21. Brito, J.G.; Roque, F.O.; Martins, R.T.; Hamada, N.; Nessimian, J.L.; Oliveira, V.C.; Hughes, R.M.; Paula, F.R.; Ferraz, S. Small forest losses degrade stream macroinvertebrate assemblages in the eastern Brazilian Amazon. *Biol. Conserv.* **2020**, *241*, 108263. [CrossRef]

22. Dala-Corte, R.B.; Melo, A.S.; Siqueira, T.; Bini, L.M.; Martins, R.T.; Cunico, A.M.; Pes, A.M.; Magalhães, A.L.; Godoy, B.S.; Leal, C.G.; et al. Thresholds of freshwater biodiversity in response to riparian vegetation loss in the Neotropical region. *J. Appl. Ecol.* **2020**, *57*, 1391–1402. [CrossRef]

23. Brejão, G.L.; Hoeinghaus, D.J.; Pérez-Mayorga, M.A.; Ferraz, S.F.B.; Casatti, L. Threshold responses of Amazonian stream fishes to timing and extent of deforestation. *Conserv. Biol.* **2018**, *32*, 860–871. [CrossRef]

24. Martins, R.T.; Brito, J.; Dias-Silva, K.; Leal, C.G.; Leitao, R.P.; Oliveira, V.C.; de Oliveira-Junior, J.M.B.; Ferraz, S.F.B.; de Paula, F.R.; Roque, F.O.; et al. Low forest-loss thresholds threaten Amazônia fish and macroinvertebrate assemblage integrity. *Ecol. Indic.* **2021**, *127*, 107773. [CrossRef]

25. Bigelow, D.P.; Borchers, A. *Major Use of Land in the United States, 2012*; EIB-176; Economic Research Service. U.S. Department of Agriculture: Washington, DC, USA, 2017.

26. Fitzpatrick, F.A.; Scudder, B.C.; Lenz, B.N.; Sullivan, D.J. Effects of multi-scale environmental characteristics on agricultural stream biota in eastern Wisconsin. *J. Am. Water Resour. Assoc.* **2001**, *37*, 1489–1507. [CrossRef]

27. Hughes, R.M.; Peck, D.V. Acquiring data for large aquatic resource surveys: The art of compromise among science, logistics, and reality. *J. N. Am. Benthol. Soc.* **2008**, *27*, 837–859. [CrossRef]

28. Hughes, R.M.; Paulsen, S.G.; Stoddard, J.L. EMAP-surface waters: A national, multiassemblage, probability survey of ecological integrity. *Hydrobiologia* **2000**, *422*, 429–443. [CrossRef]

29. Lenat, D.R.; Crawford, J.K. Effects of land use on water quality and aquatic biota of three North Carolina Piedmont streams. *Hydrobiologia* **1994**, *294*, 185–199. [CrossRef]

30. Allan, D.; Erickson, D.; Fay, J. The influence of catchment land use on stream integrity across multiple spatial scales. *Freshw. Biol.* **1997**, *37*, 149–161. [CrossRef]

31. Wang, L.; Lyons, J.; Kanehl, P.; Gatti, R. Influences of watershed land use on habitat quality and biotic integrity in Wisconsin streams. *Fisheries* **1997**, *22*, 6–12. [CrossRef]

32. Nerbonne, B.A.; Vondracek, B. Effects of local land use on physical habitat, benthic macroinvertebrates, and fish in the Whitewater River Basin, Minnesota, USA. *Environ. Manag.* **2001**, *28*, 87–99. [CrossRef] [PubMed]

33. Roth, N.E.; Allan, J.D.; Erickson, D.L. Landscape influences on stream biotic integrity assessed at multiple spatial scales. *Lands. Ecol.* **1999**, *11*, 141–156. [CrossRef]

34. Stewart, J.S.; Wang, L.; Lyons, J.; Horwatich, J.A.; Bannerman, R. Influences of watershed, riparian-corridor, and reach-scale characteristics on aquatic biota in agricultural watersheds. *J. Am. Water Resour. Assoc.* **2001**, *37*, 1475–1487. [CrossRef]

35. South, E.J.; DeWalt, R.E.; Cao, Y. Relative importance of Conservation Reserve Programs to aquatic insect biodiversity in an agricultural watershed in the Midwest, USA. *Hydrobiologia* **2019**, *829*, 323–340. [CrossRef]

36. Christensen, V.G.; Lee, K.E.; Sanocki, C.A.; Mohring, E.H.; Kiesling, R.L. *Water-Quality and Biological Characteristics and Responses to Agricultural Land Retirement in Three Streams of the Minnesota River Basin, Water Years 2006–2008*; U.S. Geological Survey: Reston, VA, USA, 2009.

37. Fore, J.; Sowa, S.P.; Galat, D.; Diamond, D.D. Assessing effects of sediment-reducing agriculture conservation practices on stream fishes. *J. Soil Water Conserv.* **2017**, *72*, 326–342. [CrossRef]

38. Lenat, D.R. Agriculture and stream water quality: A biological evaluation of erosion control practices. *Environ. Manag.* **1984**, *8*, 333–334. [CrossRef]

39. Justus, B.G.; Petersen, J.C.; Femmer, S.R.; Davis, J.V.; Wallace, J.E. A comparison of algal, macroinvertebrate, and fish assemblage indices for assessing low-level nutrient enrichment in wadeable Ozark streams. *Ecol. Indic.* **2010**, *10*, 627–638. [CrossRef]

40. Meador, M.R.; Goldstein, R.M. Assessing water quality at large geographic scales: Relations among land use, water physicochemistry, riparian condition, and fish community structure. *Environ. Manag.* **2003**, *31*, 504–517. [CrossRef] [PubMed]

41. Stauffer, J.C.; Goldstein, R.M.; Newman, R.M. Relationship of wooded riparian zones and runoff potential to fish community composition in agricultural streams. *Can. J. Fish. Aquat. Sci.* **2000**, *57*, 307–316. [CrossRef]

42. Moerke, A.H.; Lamberti, G.A. Responses in fish community structure to restoration of two Indiana streams. *N. Am. J. Fish. Manag.* **2003**, *23*, 748–759. [CrossRef]

43. Smith, D.G.; Ferrell, G.M.; Harned, D.A.; Cuffney, T.F. *A Study of the Effects of Implementing Agricultural Best Management Practices and In-Stream Restoration on Suspended Sediment, Stream Habitat, and Benthic Macroinvertebrates at Three Stream Sites in Surry County, North. Carolina, 2004–2007—Lessons Learned*; U.S. Geological Survey: Reston, VA, USA, 2011.

44. Wang, L.; Lyons, J.; Kanehl, P. Effects of watershed best management practices on habitat and fish in Wisconsin streams. *J. Am. Water Resour. Assoc.* **2002**, *38*, 663–680. [CrossRef]

45. Yoder, C.O.; Rankin, E.T.; Smith, M.A.; Alsdorf, B.C.; Altfader, D.J.; Boucher, C.E.; Miltner, R.J.; Mishne, D.E.; Sanders, R.E.; Thoma, R.F. Changes in fish assemblage status in Ohio's nonwadeable rivers and streams over two decades. In *Historical Changes in Large*

River Fish Assemblages of the Americas; Rinne, J.N., Hughes, R.M., Calamusso, B., Eds.; American Fisheries Society: Bethesda, MD, USA, 2005; pp. 399–429.

46. Effert-Fanta, E.L.; Fischer, R.U.; Wahl, D.H. Effects of riparian forest buffers and agricultural land use on macroinvertebrate and fish community structure. *Hydrobiologia* **2019**, *841*, 45–64. [CrossRef]

47. Teels, B.M.; Rewa, C.A.; Myers, J. Aquatic condition response to riparian buffer establishment. *Wildl. Soc. Bull.* **2006**, *34*, 927–935. [CrossRef]

48. Muenz, T.K.; Golladay, S.W.; Vellidis, G.; Smith, L.L. Stream buffer effectiveness in an agriculturally influenced area, southwestern Georgia. *J. Environ. Qual.* **2006**, *35*, 1924–1939. [CrossRef]

49. Lyons, J.; Weigel, B.M.; Paine, L.K.; Undersander, D.J. Influence of intensive rotational grazing on bank erosion, fish habitat quality, and fish communities in southwestern Wisconsin trout streams. *J. Soil Water Conserv.* **2000**, *55*, 271–276.

50. Sovell, L.A.; Vondracek, B.; Frost, J.A.; Mumford, K.G. Impacts of rotational grazing and riparian buffers on physicochemical and biological characteristics of southeastern Minnesota, USA, streams. *Environ. Manag.* **2000**, *26*, 629–641. [CrossRef] [PubMed]

51. Cao, Y.; Larsen, D.P.; Hughes, R.M.; Angermeier, P.L.; Patton, T.M. Sampling effort affects multivariate comparisons of stream communities. *J. N. Am. Benthol. Soc.* **2002**, *21*, 701–714. [CrossRef]

52. Silva, D.R.O.; Ligeiro, R.; Hughes, R.M.; Callisto, M. The role of physical habitat and sampling effort on estimates of benthic macroinvertebrate taxonomic richness at basin and site scales. *Environ. Monitor. Assess.* **2016**, *188*, 340. [CrossRef] [PubMed]

53. Hughes, R.M.; Herlihy, A.T.; Peck, D.V. Sampling effort for estimating fish species richness in western USA river sites. *Limnologica* **2021**, *87*, 125859. [CrossRef]

54. Whiles, M.R.; Brock, B.L.; Franzen, A.C.; Dinsmore, S.C., II. Stream invertebrate communities, water quality, and land-use patterns in an agricultural drainage basin of northeastern Nebraska, USA. *Environ. Manag.* **2000**, *26*, 563–576. [CrossRef]

55. Rinne, J.N. Effects of livestock grazing exclosures on aquatic macroinvertebrates in a montane stream in New Mexico. *Great Basin Natur.* **1998**, *48*, 146–153.

56. Herbst, D.B.; Bogan, M.T.; Roll, S.K.; Safford, H.D. Effects of livestock exclusion on in-stream habitat and benthic invertebrate assemblages in montane streams. *Freshw. Biol.* **2012**, *57*, 204–217. [CrossRef]

57. McIver, J.D.; McInnis, M.L. Cattle grazing effects on macroinvertebrates in an Oregon mountain stream. *Range Ecol. Mgmt.* **2007**, *60*, 293–303. [CrossRef]

58. Ranganath, S.C.; Hession, W.C.; Wynn, T.M. Livestock exclusion influences on riparian vegetation, channel morphology, and benthic macroinvertebrate assemblages. *J. Soil Water Conserv.* **2009**, *64*, 33–42. [CrossRef]

59. Weigel, B.M.; Lyons, J.; Paine, L.K.; Dodson, S.I.; Undersander, D.J. Using stream macroinvertebrates to compare riparian land use practices on cattle farms in southwestern Wisconsin. *J. Freshw. Ecol.* **2000**, *15*, 93–106. [CrossRef]

60. Magner, J.A.; Vondracek, B.; Brooks, K.N. Grazed riparian management and stream channel response in southeastern Minnesota (USA) streams. *Environ. Manag.* **2008**, *42*, 377–390. [CrossRef] [PubMed]

61. Bayley, P.B.; Li, H.W. Stream fish responses to grazing exclosures. *N. Am. J. Fish. Manag.* **2008**, *28*, 135–147. [CrossRef]

62. Knapp, R.A.; Matthews, K.R. Livestock grazing, Golden Trout, and streams in the Golden Trout Wilderness, California: Impacts and management implications. *N. Am. J. Fish. Mgmt.* **1996**, *16*, 805–820. [CrossRef]

63. Bowers, W.; Hosford, B.; Oakley, A.; Bond, C. *Wildlife Habitats in Managed Rangelands—The Great Basin of Southeastern Oregon*; U.S. Forest Service, University of Nebraska: Lincoln, NB, USA, 1974.

64. Keller, C.R.; Burnham, K.P. Riparian fencing, grazing, and trout habitat preference on Summit Creek, Idaho. *N. Am. J. Fish. Manag.* **1982**, *2*, 53–59. [CrossRef]

65. Stuber, R.J. Trout habitat, abundance, and fishing opportunities in fenced vs. unfenced riparian habitat along Sheep Creek, Colorado. In *Riparian Ecosystems and Their Management: Reconciling Conflicting Uses*; Johnson, R.R., Ziebell, C.D., Patton, D.R., Ffolliott, P.F., Hamre, F.H., Eds.; General Technical Report RM-120; U.S. Forest Service: Fort Collins, CO, USA, 1985; pp. 310–314.

66. Malcolm, J.W.; Radke, W.R. Effects of riparian and wetland restoration on an avian community in southeast Arizona, USA. *Open Conserv. Biol. J.* **2008**, *2*, 30–36. [CrossRef]

67. Dauwalter, D.C.; Fesenmyer, K.A.; Miller, S.W.; Porter, T. Response of riparian vegetation, instream habitat, and aquatic biota to riparian grazing exclosures. *N. Am. J. Fish. Mgmt.* **2018**, *38*, 1187–1200. [CrossRef]

68. Earnst, S.L.; Ballard, J.A.; Dobkin, D.S. Riparian songbird abundance a decade after cattle removal on Hart Mountain and Sheldon National Wildlife Refuges. In *Bird Conservation Implementation and Integration in the Americas*; Ralph, C.J., Rich, T.D., Eds.; Gen. Tech. Rep. PSW-GTR-191; U.S. Forest Service: Albany, CA, USA, 2005; pp. 550–558.

69. Poessel, S.A.; Hagar, J.C.; Haggerty, P.K.; Katzner, T.E. Removal of cattle grazing correlates with increases in vegetation productivity and in abundance of imperiled breeding birds. *Biol. Conserv.* **2020**, *241*. [CrossRef]

70. Tewksbury, J.J.; Black, A.E.; Nur, N.; Saab, V.A.; Logan, B.D.; Dobkin, D.S. Effects of anthropogenic fragmentation and livestock grazing on western riparian bird communities. *Stud. Avian Biol.* **2002**, *24*, 158–202.

71. Taylor, D.M. Effects of cattle grazing on passerine birds nesting in riparian habitat. *J. Range Mgmt.* **1986**, *39*, 254–258. [CrossRef]

72. Dobkin, D.S.; Rich, A.C.; Pyle, W.H. Habitat and avifaunal recovery from livestock grazing in a riparian meadow system of the northwestern Great Basin. *Cons. Biol.* **1998**, *12*, 209–221. [CrossRef]

73. Hurlbert, S.H. Pseudoreplication and the design of ecological field experiments. *Ecol. Monogr.* **1984**, *54*, 187–211. [CrossRef]

74. Fausch, K.D.; Torgersen, C.E.; Baxter, C.V.; Li, H.W. Landscapes to riverscapes: Bridging the gap between research and conservation of stream fishes. *BioScience* **2002**, *52*, 483–498. [CrossRef]

75. Gresswell, R.E.; Torgersen, C.E.; Bateman, D.S.; Guy, T.J.; Hendricks, S.R.; Wofford, J.E.B. A spatially explicit approach for evaluating relationships among Coastal Cutthroat Trout, habitat, and disturbance in small Oregon streams. In *Landscape Influences on Stream Habitats and Biological Assemblages*; Hughes, R.M., Wang, L., Seelbach, P.W., Eds.; American Fisheries Society: Bethesda, MD, USA, 2006; pp. 457–471.

76. Kroll, S.A.; Horwitz, R.J.; Keller, D.H.; Sweeney, B.W.; Jackson, J.K.; Perez, L.B. Large-scale protection and restoration programs aimed at protecting stream ecosystem integrity: The role of science-based goal-setting, monitoring, and data management. *Freshwat. Sci.* **2019**, *38*, 23–39. [CrossRef]

77. Kroll, S.A.; Oakland, H.C. A review of studies documenting the effects of agricultural best management practices on physiochemical and biological measures of stream ecosystem integrity. *Natur. Areas J.* **2019**, *39*, 58–77. [CrossRef]

78. O'Connor, R.J.; Walls, T.E.; Hughes, R.M. Using multiple taxonomic groups to index the ecological condition of lakes. *Environ. Monitor. Assess.* **2000**, *61*, 207–228. [CrossRef]

79. Allen, A.P.; Whittier, T.R.; Kaufmann, P.R.; Larsen, D.P.; O'Connor, R.J.; Hughes, R.M.; Stemberger, R.S.; Dixit, S.S.; Brinkhurst, R.O.; Herlihy, A.T.; et al. Concordance of taxonomic composition patterns across multiple lake assemblages: Effects of scale, body size, and land use. *Can. J. Fish. Aquat. Sci.* **1999**, *56*, 2029–2040. [CrossRef]

80. Kaufmann, P.R.; Hughes, R.M.; Whittier, T.R.; Bryce, S.A.; Paulsen, S.G. Relevance of lake physical habitat assessment indices to fish and riparian birds. *Lake Reservoir. Mgmt.* **2014**, *30*, 177–191. [CrossRef]

81. Hughes, R.M.; Bangs, B.L.; Gregory, S.V.; Scheerer, P.D.; Wildman, R.C.; Ziller, J.S. Recovery of Willamette river fish assemblages: Successes & remaining threats. In *From Catastrophe to Recovery: Stories of Fish. Management Success*; Krueger, C., Taylor, W., Youn, S.-J., Eds.; American Fisheries Society: Bethesda, MD, USA, 2019; pp. 157–184.

82. Hamilton, S.K. Biogeochemical time lags may delay responses of streams to ecological restoration. *Freshw. Biol.* **2012**, *57*, 43–57. [CrossRef]

83. Saar, D.A. Riparian livestock exclosure research in the western United States: A critique and some recommendations. *Environ. Manag.* **2002**, *30*, 516–526. [CrossRef]

84. Roni, P.; Hanson, K.; Beechie, T. Global review of the physical and biological effectiveness of stream habitat rehabilitation techniques. *N. Am. J. Fish. Manag.* **2008**, *28*, 856–890. [CrossRef]

85. Allan, J.D. Landscapes and riverscapes: The influence of land use on stream ecosystems. *Annu. Rev. Ecol. Syst.* **2004**, *35*, 257–284. [CrossRef]

86. Harding, J.S.; Benfield, E.F.; Bolstad, P.V.; Helfman, G.S.; Jones, E.B.D., III. Stream biodiversity: The ghost of land use past. *Proc. Natl. Acad. Sci. USA* **1998**, *95*, 14843–14847. [CrossRef]

87. Fesenmeyer, K.A.; Dauwalter, D.C.; Evans, C.; Allai, T. Livestock management, beaver, and climate influences in a semi-arid landscape. *PLoS ONE* **2018**, *13*. [CrossRef]

88. Van Meter, K.J.; Cappellen, P.V.; Basu, N.B. Legacy nitrogen may prevent achievement of water quality goals in the Gulf of Mexico. *Science* **2018**, *360*, 427–430. [CrossRef] [PubMed]

89. Kauffman, J.B.; Beschta, R.L.; Otting, N.; Lytjen, D. An ecological perspective of riparian and stream restoration in the western United States. *Fisheries* **1997**, *22*, 12–24. [CrossRef]

90. Hynes, H.B.N. The stream and its valley. *Verh. Internat. Verein. Limnol.* **1975**, *19*, 1–15. [CrossRef]

91. Armour, C.; Duff, D.; Elmore, W. The effects of livestock grazing on western riparian and stream ecosystem. *Fisheries* **1994**, *19*, 9–12. [CrossRef]

92. Feld, C.K.; Fernandes, M.R.; Ferreira, M.T.; Hering, D.; Ormerod, S.J.; Venohr, M.; Gutiérrez-Cánovas, C. Evaluating riparian solutions to multiple stressor problems in river ecosystems—A conceptual study. *Water Res.* **2018**, *139*, 381–394. [CrossRef]

93. Laitos, J.G.; Ruckriegle, H. The clean water act and the challenge of agricultural pollution. *Vt. Law Rev.* **2013**, *37*, 1033–1070.

94. Karr, J.R.; Dudley, D.D. Ecological perspective on water quality goals. *Environ. Manag.* **1981**, *5*, 55–68. [CrossRef]

95. Hughes, R.M.; Noss, R.F. Biological diversity and biological integrity: Current concerns for lakes and streams. *Fisheries* **1992**, *17*, 11–19. [CrossRef]

96. Brewin, M.K.; Monita, D.M.A. Forest-fish conference: Land managment practices affecting aquatic ecosystems. In Proceedings of the Forest-Fish Conference, Calgary, AB, Canada, 1–4 May 1996; pp. 13–30.

97. Davies, S.P.; Jackson, S.K. The biological condition gradient: A descriptive model for interpreting change in aquatic ecosystems. *Ecol. Appl.* **2006**, *16*, 1251–1266. [CrossRef]

98. Ohio EPA. *Biological Criteria for the Protection of Aquatic Life*; Division of Water Quality Planning and Assessment, Ohio Environmental Protection Agency: Columbus, OH, USA, 1988.

99. Bigelow, D.; Hellerstein, D. *In Recent Years, Most Expiring Land in the Conservation Reserve Program Returned to Crop Production. Economic Research Service*; U.S. Department of Agriculture: Washington, DC, USA, 2020. Available online: https://www.ers.usda.gov/amber-waves/2020/february/in-recent-years-most-expiring-land-in-the-conservation-reserve-program-returned-to-crop-production (accessed on 6 July 2021).

100. USDI (U. S. Department of the Interior). *The Federal Land Policy and Management Act of 1976*; Bureau of Land Management: Washington, DC, USA, 2016. Available online: https://www.blm.gov/sites/blm.gov/files/AboutUs_LawsandRegs_FLPMA.pdf (accessed on 6 July 2021).

101. Flynn, R. Daybreak on the land: The coming of age of the Federal Land Policy and Management Act of 1976. *Vt. Law Rev.* **2005**. Available online: https://lawreview.vermontlaw.edu/wp-content/uploads/2012/02/flynn.pdf (accessed on 6 July 2021).

102. Wood, M.C. *Nature's Trust: Environmental Law for a New Ecological Age*; Cambridge University Press: New York, NY, USA, 2014.

103. Congdon, C.H.; Young, T.F.; Gray, B.E. Economic incentives and nonpoint source pollution: A case study of California's grasslands region. *Hastings Environ. Law J.* **1995**, *2*, 185–247. Available online: https://repository.uchastings.edu/hastings_environmental_law_journal/vol2/iss3/3 (accessed on 6 July 2021).

104. Bates, S. Bridging the governance gap: Emerging strategies to integrate water and land use planning. *Nat. Resour. J.* **2012**, *52*, 61–97.

105. Gardner-Andrews, N. Water quality and land use planning: Emerging legal and regulatory considerations. *Plan. Environ. Law* **2013**, *65*, 4–9. [CrossRef]

106. Feio, M.J.; Hughes, M.; Callisto, M.; Nichols, S.; Odume, O.; Quintella, B.; Kuemmerlen, M.; Aguiar, F.; Almeida, S.; Alonso-Eguíalis, P.; et al. The Biological Assessment and Rehabilitation of the World's Rivers: An Overview. *Water* **2021**, *13*, 371. [CrossRef]

107. Flávio, H.M.; Ferreira, P.; Formigo, N.; Svendsen, J.C. Reconciling agriculture and stream restoration in Europe: A review relating to the EU Water Framework Directive. *Sci. Tot. Environ.* **2017**, *596–597*, 378–395. [CrossRef] [PubMed]

108. Hughes, R.M.; Infante, D.M.; Wang, L.; Chen, K.; Terra, B.F. *Advances in Understanding Landscape Influences on Freshwater Habitats and Biological Assemblages*; American Fisheries Society: Bethesda, MD, USA, 2019.

109. Marsh, P.C.; Luey, J.E. Oases for aquatic life within agricultural watersheds. *Fisheries* **1982**, *7*, 16–24. [CrossRef]

110. Beschta, R.L.; Ripple, W.J. Riparian vegetation recovery in Yellowstone: The first two decades after wolf reintroduction. *Biol. Conserv.* **2016**, *198*, 93–103. [CrossRef]

MDPI

St. Alban-Anlage 66

4052 Basel

Switzerland

Tel. +41 61 683 77 34

Fax +41 61 302 89 18

www.mdpi.com

Water Editorial Office

E-mail: water@mdpi.com

www.mdpi.com/journal/water